Graphene–Electrolyte Interfaces

Graphene-Electrolyte Interfaces

Electronic Properties and Applications

Hualin Zhan

CRC Press
Taylor & Francis Group
Boca Raton London New York

CRC Press is an imprint of the
Taylor & Francis Group, an **informa** business

Graphene–Electrolyte Interfaces: Electronic Properties and Applications

First published 2020 by Jenny Stanford Publishing Pte. Ltd.

Published 2023 by CRC Press
Taylor & Francis Group
6000 Broken Sound Parkway NW, Suite 300
Boca Raton, FL 33487-2742

ISBN 978-981-4774-94-9 (hbk)
ISBN 978-1-003-04487-1 (ebk)

Visit the Taylor & Francis Web site at
http://www.taylorandfrancis.com

and the CRC Press Web site at
http://www.crcpress.com

British Library Cataloguing-in-Publication Data
A catalogue record for this book is available from the British Library.

Contents

Preface

Shortly after the Graphene Conference held in Barcelona, Spain, at the end of March 2017, I received an invitation from Jenny Stanford Publishing to write a book on "molecular detection in liquids by graphene Hall effect measurement," the topic of my talk at the conference. The sensitivity of the measurement to the molecules within a liquid shown in the presentation indicated that much could be done in this direction (now, though, I see more papers published on this topic after the conference). However, I started to wonder whether the fundamentals of this topic could be helpful for other research projects.

Discovered in 2004, graphene is a very exciting two-dimensional material. After many of its unique properties were reported in the recent decade, the science and engineering community began to apply it in different systems for various purposes. Graphene–electrolyte interfacial processes are among the most interesting topics for applications in biology, electronics, materials engineering, chemical engineering, energy storage, and electrochemistry. The theoretical and experimental techniques used in the study of the liquid gated Hall measurement of graphene are essential to understand the graphene–electrolyte systems. Therefore, I decided to emphasize these systems and techniques in the book.

It has been a long and interesting journey to study graphene nanoelectronics and electrochemistry. I first learned the physics and applications of low-dimensional semiconductors in the subject of nanoelectronics taught by Prof. Thomas Szkopek. He explained the physics of low-dimensional semiconductors in an elegant fashion which every student could understand. It was a privilege, and I am still very grateful for his introduction, as it turned out to be very helpful for the research I did later. My adventure in electrochemistry began after my supervisor left the university during my PhD study. I then started to work with an experienced experimental electrochemist, Dr. David J. Garrett, because of whom I became particularly interested in semiconductor electrochemistry and photoelectrochemistry. However, it was difficult to find a reference that described the fundamental theoretical problems of electrons within graphene–electrolyte systems, in spite of many experimental works reported for various application. Hence, I believed that it would be convenient for the readers if I summarized the existing understanding related to this field in this book. I must thank Jenny Stanford Publishing for offering this opportunity to me to share the knowledge. I would also like to thank Mr. Alexander Tsai for the help in checking the grammar in chapters 1 and 2 of the book.

Hualin Zhan
The University of Melbourne
Australia, December 2017

Chapter 1

Introduction

Long before the gold rush of research in two-dimensional (2D) materials from 2004, the study of low-dimensional electronics developed in 20th century has delivered a deeper understanding of semiconductor physics to the science and engineering community [Davies (1998)]. The knowledge obtained has revolutionized the electronic technology and greatly improved the quality of life. From the first invention of personal computers to the popular use of pocket-sized smartphones, people intend to design faster electronic devices with smaller sizes. Graphene, as an iconic 2D material discovered by K. S. Novoselov and A. K. Geim in 2004 [Novoselov et al. (2004)], has an atomic thickness, ultrafast electrons, high flexibility, mechanical robustness, and more. This material fits ideally to the development trend of electronic devices, and it was then believed to hold the future for the next generation of electronics [Semiconductor Industry Association (2009)].

Liquid, termed as solution or electrolyte in some cases, has played a central role in the studies of chemistry and biology. This is because most of the chemical processes of interest in these fields occur in the liquid phase. As the information for the chemical processes could be difficult to visualize, it is useful and important for people to be able to quantitatively investigate these processes

Graphene–Electrolyte Interfaces: Electronic Properties and Applications
Hualin Zhan

ISBN 978-981-4774-94-9 (Hardcover), 978-1-003-04487-1 (eBook)
www.jennystanford.com

in analytical chemistry through other means, such as electrical and optical measurements [Bard and Faulkner (2000)]. Due to its intriguing electrical properties, graphene is among the perfect candidates for probing or even influencing the chemical/physical processes in liquid. Conversely, unique properties (including an electrical double layer, a high dielectric constant, etc.) offered by electrolytes which contain mobile ions can be used to enhance the performance of the graphene electronic devices. This reciprocal relation suggests that there is a promising future in the research of graphene–electrolyte systems, where the electronic and ionic properties still require systematic studies to be understood.

In this book, we first focus on the fundamental theory to understand the electronic properties at the graphene–electrolyte interface, then review the common techniques for graphene and electrolyte, and finally discuss the relevant applications. This is because electrons are inevitably involved in the interaction between graphene and electrolyte, and the applications in biology and electronics involving the electronic activities in graphene–electrolyte systems are becoming increasingly important in the global community. Ionic properties in the systems, which are useful for applications such as energy storage, are outside the scope of this book. Note that the solid-state ionic conductors, sometimes referred to as solid electrolytes, are not discussed in this book.

1.1 Graphene

Graphene is a free-standing monoatomic layer of graphite, where the carbon atoms are arranged in a honeycomb structure (Fig. 3.1a). Since the initial experimental proof of graphene's existence in 2004, graphene has drawn much attention due to its atomic and electronic 2D nature. Despite the widely held belief that graphene was thermodynamically unstable, its fabrication opened an era for the research in other atomic 2D thin films [Geim and Novoselov (2007)], its peculiar mechanical [Lee et al. (2008, 2013a)], electrical [Castro Neto et al. (2009); Das Sarma et al. (2011); Novoselov et al. (2004)], optical [Crassee et al. (2010); Nair et al. (2008)] and magnetic properties [Gu et al. (2011); Nomura and MacDonald

(2006); Novoselov et al. (2005); Ping et al. (2014); Sun et al. (2010)] have stimulated the research in various fields, such as nanoelectronics [Britnell et al. (2012); Farmer et al. (2009); Georgiou et al. (2013); Lin et al. (2010, 2009); Schwierz (2010); Wu et al. (2011); Xia et al. (2010a)], plasmonics [Abajo (2013); Buljan et al. (2013); Carminati (2013); Crassee et al. (2012); Fei et al. (2013); Grigorenko et al. (2012); Hoggard et al. (2013); Ju et al. (2011); Li et al. (2014b); Xia et al. (2013); Yan et al. (2012, 2013)], optics [Liu et al. (2011)], magnetics [Côté et al. (2011); Lu et al. (2011)], electrochemistry [Ambrosi et al. (2014); Brownson et al. (2012); El-Kady et al. (2012); McCreery (2008); Pumera (2013); Xiong et al. (2014)], and chemical/biological sensing [Ahmed et al. (2012); Antony and Grimme (2008); Berland et al. (2011); Chakarova-Käck et al. (2006); Garaj et al. (2010); Gowtham et al. (2007); Kim et al. (2010); Kostarelos and Novoselov (2014a,b); Kumar et al. (2013); Le et al. (2012); Liu et al. (2012); Min et al. (2011); Ortmann et al. (2005); Pumera et al. (2010); Schedin et al. (2007); Shang et al. (2008); Shao et al. (2010); Siwy and Davenport (2010); Traversi et al. (2013); Wang et al. (2009); Zhu et al. (2013)] and so on. The biological applications of graphene include drug delivery, biosensing, bionic implants [Hess et al. (2013)], and others [Ferrari et al. (2015); Nature Nanotechnology Feature (2014)].

In essence, the unique electronic properties of graphene as a material are the key to understanding the fascinating behavior seen not only electronically, but also optically and magnetically. As graphene is categorized as a "semiconductor" on some occasions, the current knowledge of low-dimensional semiconductors can be adopted to understand the electronic behavior in graphene. Therefore, this book first reviews the physics of semiconductors in Chapter 2, and then applies this knowledge to graphene (Chapter 3), to build the foundational understanding of its electronic behavior. The most important fundamental concepts are the density of states and the Fermi distribution function, as they are used to determine both electron density and quantum capacitance in graphene. Mobility is also an essential and useful parameter when discussing the performance of electronic devices, for example, it could be included in the Drude model for the calculation of conductivity. When graphene is incorporated with other materials (or electrolytes) during the fabrication or operation of a device, the electronic behavior may further change. Therefore, a discussion of heterostructures is also included.

In general, graphene devices involving the use of electrolyte are not operated in harsh environment (e.g., ultralow temperature, super-high magnetic field, nanoscale structure, high electric voltage), suggesting that some of the quantum behaviors are not observable, e.g., the Shubnikov–de Haas oscillation [Novoselov et al. (2005)] and Klein tunneling [Castro Neto et al. (2009)]. Classical theories, including the Boltzmann transport theory (appendix A) and the Drude model, are used to describe the electron transport, while quantum mechanics is used to derive the electron density due to the crystalline structure of graphene.

The experimental methods for graphene, including the growth techniques, characterizations, and device fabrication, are then briefly summarized in Chapter 6 and appendix F. This summary not only introduces common experimental techniques to the readers, but it also helps to deepen the understanding of electron transport in graphene.

1.2 Electrolyte

Electrolyte, a liquid containing mobile ions, usually attracts more attention in chemical and life sciences than in physics as it serves as a platform upon which most chemical reactions of interest take place. Electrochemical processes, which are accompanied by an electrical current between different chemical phases (e.g., solid electrodes and liquid electrolytes), are particularly attractive for the science and engineering community [Bard and Faulkner (2000)]. This is because we are able to intervene or monitor the chemical reactions or ionic activities (for example, in electrolyte) by applying an external electrical potential. Electrochemistry is therefore a field which studies the relation between electrical and chemical effects, and the corresponding techniques are often referred as electrochemical methods. Hence, it is important to review the theoretical and experimental techniques of electrochemistry to properly study the electrical properties at a graphene–electrolyte interface.

If a reduction or oxidation occurs in an electrolyte due to the charge transfer between the electrode–electrolyte interface, the process is referred to as being Faradaic. However, if the ionic structure changes at the electrode–electrolyte interface with no occurrence of chemical reactions in response to the applied electrical potential, the process is then non-Faradaic. Note that a current could still be observed in non-Faradaic process when applying a (time-dependent) voltage. This current is called displacement current or capacitive current. There are several different but related fundamental theories for Faradaic and non-Faradaic processes. These theoretical models are briefly introduced in Chapter 4.

Popular applications of experimental electrochemical methods include electrolysis, electroplating, electroetching, biosensing, neuron stimulation, energy storage, among many others. The electrochemical processes for electrolysis, electroplating, and electroetching are mainly Faradaic, while biosensing, neuron stimulation, and energy storage may consist of both Faradaic and non-Faradaic processes. On some occasions, graphene is simply considered as an electrode in graphene–electrolyte systems, where observed electrical current includes both Faradaic and non-Faradaic components (e.g., amperometric biosensors or pseudocapacitors). It is therefore useful to review classical electrochemical methods for electrolyte (Chapter 7).

1.3 Graphene–Electrolyte Systems

Since graphene is an electron/hole conductor while electrolyte is an ionic conductor, the investigation of the electrical processes must be taken with care. As electrons are transferred between the ions in electrolyte and graphene, the electrical current of the Faradaic process can still be described by the model in section 4.2. Since graphene is very sensitive to the environment due to its 2D nature, additional spontaneous charge transfer processes (or doping) could happen. This is described in section 5.1.1. On the other hand, the electrical double layer capacitance (caused by a non-Faradaic process) at the electrode–electrolyte interface and the dielectric constant of the electrolyte could also affect the electronic behavior in graphene, as will be discussed in the remainder of Chapter 5.

Based on the theoretical understanding behind the electronic properties of graphene–electrolyte interfaces in Chapter 5 and the

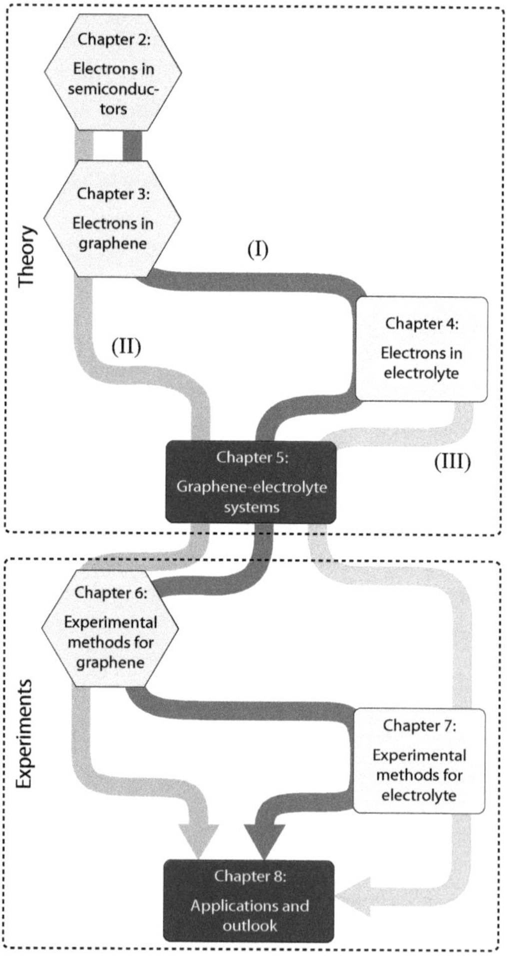

Figure 1.1 Suggested approaches for the readers with different background. (I) Readers who never worked with graphene nanoelectronics and electrochemistry. (II) Readers who are very familiar with theoretical and experimental electrochemistry. (III) Readers who have experience in graphene nanoelectronics.

experimental techniques summarized in Chapters 6 and 7, several current applications of graphene devices involving electrolyte are reviewed in Chapter 8. As the ultrasensitivity which the electrons in graphene have to the environment can be ideal for some biological applications, devices for biosensing and neural activity recording which take the advantage of the electronic properties at graphene–electrolyte interfaces are described. In addition, an example of enhancing a graphene nanoelectronic device is given in the same chapter. At the end of the book, the outlook for future studies of graphene–electrolyte systems is presented.

This book introduces the fundamental theories for the electronic processes at graphene–electrolyte interfaces and then discusses the applications. It can be used by the readers from different fields of research including physicists, chemists, biologists, and engineers. Due to the variety of background knowledge readers may have, there are three approaches suggested in Fig. 1.1 for using this book more efficiently. Readers who have never worked with graphene nanoelectronics and electrochemistry are highly recommended to go through all the chapters in this book (approach I). As some of the important concepts in graphene–electrolyte systems are not used frequently even in some research projects involving graphene or electrolyte, for example, the quantum capacitance (sections 2.2.3.2 and 3.2.2) and the Gerischer model (section 4.1.3). Even readers who have experience in conventional graphene electronics and electrochemistry are also recommended to follow approach I. However, if the readers are very familiar with both theoretical and experimental electrochemistry, Chapters 4 and 7 could be skipped (approach II). For the readers who are experts in graphene nanoelectronics (both theory and experiments), approach III is recommended.

Chapter 2

Electrons in Semiconductors

Electron transport phenomena in solid materials have been of great interest to physicists since the late nineteenth century when J. J. Thomson discovered electrons. Construction of a systematic theory for electrical conductivity therefore became necessary. During the course of this search, different models appeared, some of which successfully explained the experimental results in a qualitative way. However, the failure to quantitatively analyze experiments using these theories stimulated further investigation. It is worth noting that although the early theories for electron transport are no longer valid under certain conditions, they are still of importance when used appropriately.

The first theory constructed by P. Drude in 1900 studied the excellent electrical conduction in metals where electrons are considered as collisional charged particles, as metals occupy around two thirds of elements in the periodic table. The Drude model has also shown an exceptional success in describing the electrical conductivity in semiconductors. Semiconductor is a material whose electrical conductivity falls between that of a metal and an insulator, which could be controlled externally (such as temperature, electric field, light illumination, etc.). While Drude model and its more sophisticated companion, Boltzmann transport theory, are useful for predicting the classical electrical conductivity and some plasmonic

Graphene–Electrolyte Interfaces: Electronic Properties and Applications
Hualin Zhan

ISBN 978-981-4774-94-9 (Hardcover), 978-1-003-04487-1 (eBook)
www.jennystanford.com

behavior, the conductivity transition in semiconductors, among many other unconventional nanoscale phenomena, is bettered described by later developed quantum theory.

In this chapter, we will briefly review both the classical and quantum theories for electron transport in semiconductors developed in the last century, as well as their modern electrical applications. These theories are key to understanding the electronic behavior in graphene, which is considered a special semiconductor (semimetal). Due to the similarities between conventional semiconductors and graphene in theory, current electronic applications of semiconductors provide research directions for graphene devices, including graphene–electrolyte systems.

2.1 Free Electron Gas

The first model proposed by Drude at the turn of the twentieth century describes the electrical conduction of metals based on the kinetic theory of gases, i.e., the formation of electron gas. Electrons can move freely (in straight lines) in the gas until a collision event occurs. This is similar to the motion of molecules in gas, and therefore this theory can be described as the **free electron gas** model or the Drude model.

2.1.1 The Drude Model

Drude's theory of electron transport in solids can be developed using classical statistical mechanics based on the Boltzmann transport equation (BTE) for the kinetic theory of gases.

For a system with a large number of particles (such as a gas), it is of interest to investigate the collective behavior, as opposed to focusing in on the singular particle. In most cases, we want to find the density of the particles and predict their motion. Phase space (position-momentum/velocity space) is a powerful concept proposed by mathematicians and physicists including J. Liouville, L. Boltzmann, and others in the late nineteenth century [Nolte (2010)], in which all possible states of a dynamical system are represented.

In classical kinetic theory, the distribution function of particles ($f = f(\vec{r}, \vec{v}, t)$, where $\vec{r}$, $\vec{v}$, t are position, velocity, and time, respectively) is defined as their number density in phase space. The famous **transport equation** derived by Boltzmann in 1876 describes the evolution of the particle distribution function as a function of time, and is written as

$$\frac{\mathrm{d}f}{\mathrm{d}t} = \frac{\partial f}{\partial t} + \vec{v} \cdot \nabla f + \frac{\vec{F}}{m} \cdot \nabla_v f = \left(\frac{\partial f}{\partial t}\right)_c, \tag{2.1}$$

where $\vec{F}$ is the applied external force, m is the mass of a particle, and $\left(\frac{\partial f}{\partial t}\right)_c$ is a collision term which represents the perturbation of the distribution function by collisions. $\nabla = \hat{e}_x \frac{\partial}{\partial x} + \hat{e}_y \frac{\partial}{\partial y} + \hat{e}_z \frac{\partial}{\partial z}$ and $\nabla_v = \hat{e}_x \frac{\partial}{\partial v_x} + \hat{e}_y \frac{\partial}{\partial v_y} + \hat{e}_z \frac{\partial}{\partial v_z}$ are the gradient operators in position and velocity spaces, respectively. $\hat{e}_x$, $\hat{e}_y$, and $\hat{e}_z$ are the standard basis vectors for a three-dimensional Cartesian coordinate system with coordinates (x, y, z). This is ideal for the description of non-equilibrium systems.

In general, it is still impossible to obtain the global analytical solution of Eq. 2.1 without any assumptions at this stage, though a recent study has proven the existence of a solution for long-range interactions [Gressman and Strain (2010)]. Therefore, the following assumptions are made for the BTE when investigating electron transport in solid materials using the Drude model.

1. *The distribution function is Maxwell–Boltzmann.* In statistical mechanics, classical particles which are described by Newton's laws are distinguishable from each other, and their trajectories are traceable. The Pauli exclusion principle does not apply to these particles, and therefore their population is described by a Maxwell–Boltzmann distribution.

2. *An electron, as a classical particle, is only scattered by the atoms or ions in the material.* The time taken by a collision process is negligible and the trajectory of an electron is unaffected between two collision events.

3. *Electron–electron interaction is ignored.* This is because in statistical mechanics the distribution function of particles describes the probable state of the system. In other words, the population of electrons at each energy level (i.e., the distribution function) already accounts for electron–electron interaction.

In order to obtain useful solutions for a practical physics problem, it is convenient to integrate the BTE over different orders of momentum (velocity) [Boyd and Sanderson (2003); Zhan and Hu (2011); Zhan et al. (2011a)]. This technique is defined as taking the moments of BTE (refer to appendix A for more details). With the previously stated approximations, it is then straightforward to obtain **the Drude model** by taking the first order moment

$$m\frac{\mathrm{d}\vec{u}}{\mathrm{d}t} = \vec{F} - \frac{m\vec{u}}{\tau}, \tag{2.2}$$

where $\vec{u}$ is the mean velocity of the electrons and τ is the relaxation time (the average time interval between two collision events). $\frac{m\vec{u}}{\tau}$ represents the "friction" of the particles due to collision. This term will be discussed as follows.

2.1.1.1 Electron scattering and mobility

In contrast to the example shown in appendix A, where $\left(\frac{\partial f}{\partial t}\right)_c = 0$, the collision term $\left(\frac{\partial f}{\partial t}\right)_c$ must be non-zero when studying the charge transport in solid materials. This is because charge transport processes are often collisional and the distribution function changes in the presence of an external perturbation, such as an electric field. A commonly used model for the collision term is the BGK (Bhatnagar–Gross–Krook) operator

$$\left(\frac{\partial f}{\partial t}\right)_c = -\frac{f - f_0}{\tau}, \tag{2.3}$$

where $f = f_0 + f_1$. f_0 and f_1 represent the equilibrium term and the perturbation term of the distribution function, respectively. This is a coarse approximation of the collision process stating that the change of the distribution function is linear and happens instantly after a duration (relaxation time τ). As the collision process is simply described using the relaxation time, Eq. 2.3 is also called the **relaxation time approximation**.

By substituting Eq. 2.3 into Eq. 2.1 and taking the first-order moment, we are able to obtain Eq. 2.2. The mathematical detail is not shown here but readers are encouraged to perform the derivation by referring to appendix A.3.2. Note that a relation of material derivative $\frac{\mathrm{d}\vec{u}}{\mathrm{d}t} = \left(\frac{\partial}{\partial t} + \vec{u}\cdot\nabla\right)\vec{u}$ from fluid mechanics is used here.

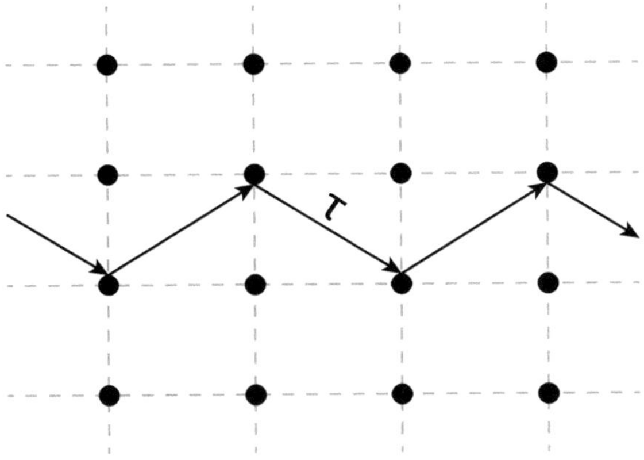

Figure 2.1 Trajectory of an electron (black lines with arrows) in a solid material, where the black spheres represent the atoms or ions and τ is the relaxation time.

$\frac{\mathrm{d}f_0}{\mathrm{d}t} = 0$, as the equilibrium part of the distribution function does not change with time.

Equation 2.2 describes the motion of an electron in a solid material, where the left-hand side (LHS) denotes the overall acceleration (force) on the electron, $\vec{F}$ at the right-hand side (RHS) is the applied external force, and $\frac{m\vec{u}}{\tau}$ is the friction due to collision. The physical image of the collision process is illustrated in Fig. 2.1.

In the presence of an electric field, the force on an electron in a conductor is $\vec{F} = q\vec{E}$, where $q = -e = -1.60217662 \times 10^{-19}$ C for an electron and $\vec{E}$ is the electric field strength. At equilibrium the mean velocity of electrons in the conductor reaches a stable value resulting in the LHS of Eq. 2.2 equaling 0. Hence,

$$\vec{F} = q\vec{E} = \frac{m\vec{u}}{\tau}. \tag{2.4}$$

Rearrangement gives

$$\vec{u} = \frac{q\tau}{m}\vec{E} = \mu\vec{E}, \tag{2.5}$$

where $\mu = \frac{q\tau}{m}$ is the **mobility** of the electron. This is a useful concept in solid state physics which describes how fast an electron moves in response to an external electric field, and it can be used to estimate the performance of an electronic device.

2.1.1.2 DC electrical conductivity

The DC electric current density ($\vec{J}$) is defined as the electric current per area. $\vec{J}$ in a conductor is written as

$$\vec{J} = \sigma \vec{E} = qn\vec{u}, \tag{2.6}$$

where σ is the **electrical conductivity**. Substitution of Eq. 2.5 into Eq. 2.6 yields

$$\sigma = qn\mu. \tag{2.7}$$

2.1.2 Fermi–Dirac Distribution

L. de Broglie postulated the revolutionary idea of a matter wave in his PhD thesis in 1924, which serves as one of the foundation stones for modern quantum mechanics. It states that a particle is a wave at the same time, and the wavelength is described using a relation $\lambda = \frac{h}{p}$, where $h = 6.62607004 \times 10^{-34}\ \mathrm{m^2 kg\, s^{-1}}$ is Planck's constant and p is the momentum. Therefore, an electron in a solid material can be thought of as a wave. This concept serves as one of the cornerstones for solid state physics, semiconductor electrochemistry, along with many fields.

The **quantum state** is a concept used to describe the state of an isolated quantum object (e.g., an electron), where the observables (such as momentum) could be predicted. In the context of atomic physics, the state (or state vector) consists of a set of variables including principal quantum number, angular momentum quantum number, magnetic quantum number, and spin z-component. Here in solid-state physics, we start from considering the momentum (and position) of electrons for the quantum state, where spin component could be added later (e.g., g in Eq. 2.19). As electrons with various wavelengths have different momenta according to de Broglie's theorem, the number of waves with different wavelengths can be used as an indication of the number of quantum states.

For a region in a one-dimensional phase space with length x_1 and momentum p_1, the largest wavelength of a particle is the same as the space dimension ($\lambda = x_1$). In other words, the smallest non-zero momentum of a particle is $\frac{h}{x_1}$. Therefore, the number of particles

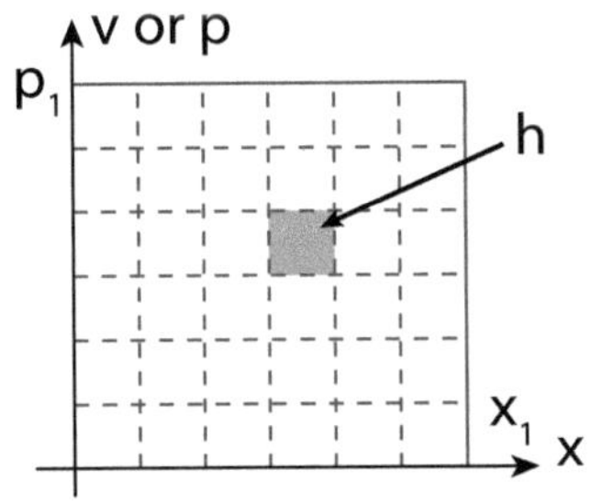

Figure 2.2 Illustration of a region (defined by solid lines) in one-dimensional phase space, where x_1 and p_1 are the maximum length and the momentum. The dashed grid demonstrates the quantum states, in which the area of the smallest division (a quantum state) is h.

with different wavelengths in the region is

$$N_D = \frac{p_1}{h/x_1} = \frac{V_p}{h}, \tag{2.8}$$

where $V_p = p_1 x_1$ is the area of the region in phase space. N_D is also called the **number of quantum states** (or the degree of degeneracy, if these states share the same energy), and hence *h represents the area of a single quantum state in one-dimensional phase space*, as illustrated in Fig. 2.2. In a phase space with the dimensionality r, h^r is the volume of a single quantum state.

In experiments, we are often interested in knowing the number (or density) of particles in a system. Although Eq. 2.8 describes the number of quantum states in the system, it is unclear how many particles are there in each of the quantum states. The famous **Pauli exclusion principle** was formulated by W. Pauli in 1925, which states that *there is no more than one electron in a quantum state* (with volume of h^r in phase space). This theorem is later extended to the systems consisting of identical fermions (particles with half-integer spins), of which electrons are a part of. However, this still does not give a statistical result of the population of particles in a system.

A year later, E. Fermi and P. Dirac independently discovered a mathematical method to describe the statistics of fermions based on the Pauli exclusion principle and an idea that all identical fermions are indistinguishable from each other [Pathria and Beale (2011)].

This **Fermi–Dirac distribution** function can be written

$$f_{\mathrm{FD}}(\mathcal{E}) = \frac{1}{e^{\frac{\mathcal{E}-\mu_{\mathrm{p}}}{k_{\mathrm{B}}T}} + 1}, \tag{2.9}$$

where $\mathcal{E}$ is the energy, μ_{p} is the chemical potential, $k_{\mathrm{B}} = 1.38064852 \times 10^{-23}\,\mathrm{m^2kg\,s^{-2}K^{-1}}$ is the Boltzmann constant, and T is the temperature in Kelvin.

In contrast to Maxwell–Boltzmann distribution function which directly describes the number density of particles in phase space, *Fermi–Dirac statistics represents the occupation probability of a quantum state by a fermion (electron) in energy space,* due to its derivation process [Pathria and Beale (2011)] where

$$N = \int f_{\mathrm{FD}}(\mathcal{E})D(\mathcal{E})\mathrm{d}\mathcal{E}. \tag{2.10}$$

N is the total number of particles and $D(\mathcal{E})$ is the **density of quantum states** (number of quantum states per energy, which will be discussed in section 2.1.3.3).

Figure 2.3a shows a Fermi–Dirac distribution function ($\mu_{\mathrm{p}} = 1.24$ eV) at room temperature (solid line). The occupation probability at low energy is 1 but at higher energy drops to 0. A smoothed step transition near μ_{p} is observed, where $f = 0.5$. When the temperature decreases, the transition becomes sharper, as illustrated by the dotted line in Fig. 2.3a. It is worth noting that the shape of this distribution function does not change too much from the room temperature distribution. It is then common to estimate electron density using the Fermi–Dirac distribution function when $T = 0$ K in solid state physics. The chemical potential in this case is defined as the **Fermi level** ($\mathcal{E}_{\mathrm{F}} = \mu_{\mathrm{p}}(T = 0\ \mathrm{K})$).

The Maxwell–Boltzmann distribution function for the energy is illustrated in Fig. 2.3b (dashed line), where its normalized form is written

$$f_{\mathrm{MB}}(\mathcal{E}) = 2\sqrt{\frac{\mathcal{E}}{\pi}}\left(\frac{1}{k_{\mathrm{B}}T}\right)^{\frac{3}{2}} \exp\left(-\frac{\mathcal{E}}{k_{\mathrm{B}}T}\right). \tag{2.11}$$

Note that this equation is derived from the normalized Maxwell-Boltzmann distribution in velocity space ($f(\vec{v})$) by letting $1 = \int_0^{\infty} f_{\mathrm{MB}}(\mathcal{E})\mathrm{d}\mathcal{E} = \int_{-\infty}^{\infty} f_{\mathrm{MB}}(\vec{v})\mathrm{d}^3\vec{v} = \int_0^{\infty} 4\pi v^2 f_{\mathrm{MB}}(v)\mathrm{d}v$. The comparison of f_{MB} and f_{FD} in linear and logarithmic scales are shown in

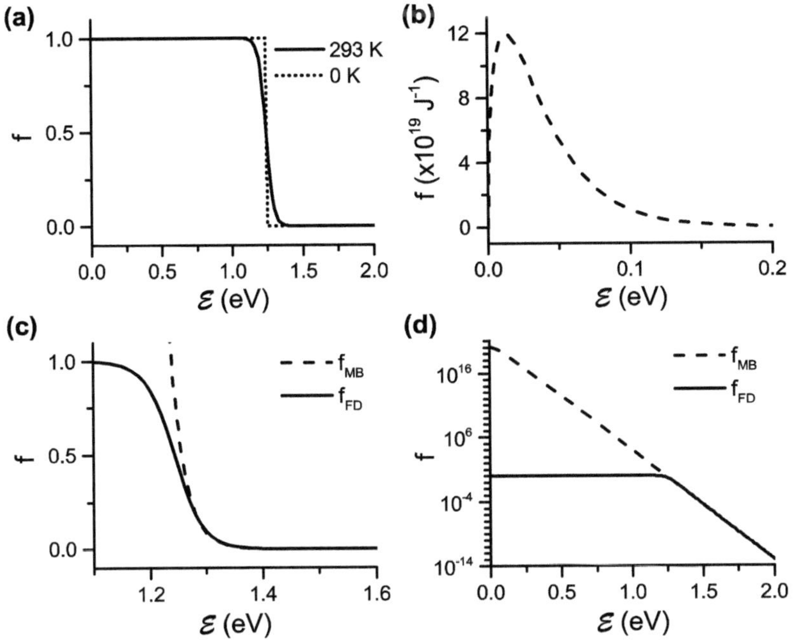

Figure 2.3 (a) Fermi–Dirac distribution for $T = 293$ K (solid) and $T = 0$ K (dotted) where the chemical potential $\mu_\mathrm{p} = 1.24$ eV. (b) Maxwell–Boltzmann distribution for the energy. (c) Comparison of Fermi–Dirac and Maxwell–Boltzmann distribution functions in linear scale near the chemical potential. (d) Comparison of Fermi–Dirac and Maxwell–Boltzmann distribution functions in logarithmic scale.

Figs. 2.3c–d, respectively. It is seen that f_MB *and* f_FD *are the same at higher energy (when* $\mathcal{E} \gg \mu_\mathrm{p}$*).*

The Fermi–Dirac distribution function was later applied to BTE by A. Sommerfeld to develop a semiclassical theory, where the results for electrical conduction are very similar to the Drude model. Readers who are interested in the details of this theory are recommended to read solid state physics textbooks [Ashcroft and Mermin (1976); Kittel (2004)]. There are still some fundamental questions which cannot be answered using this model. For example, what determines the number of conduction electrons, and why are some materials nonconductive? It is therefore important to study the mechanics of electrons at the quantum level.

2.1.3 Quantum Mechanics

Apart from the influence of de Broglie's matter–wave concept on the previously discussed statistics, the matter–wave concept inspired the development of mechanical equations for particles in a very small scale in the early twentieth century. E. Schrödinger derived a wave equation to describe the matter wave in 1925 and later published it in 1926. The **Schrödinger's equation** was widely accepted after its application in reproduction of the Bohr model formula for the energy levels. It can be written as

$$\left[-\frac{\hbar^2}{2m}\nabla^2 + V(\vec{r})\right]\Psi(\vec{r}, t) = i\hbar\frac{\partial\Psi(\vec{r}, t)}{\partial t}, \tag{2.12}$$

where $\hbar = \frac{h}{2\pi}$, i is the imaginary unit, $V(\vec{r})$ is a potential energy function, and $\Psi(\vec{r}, t)$ is the wave function which is dependent on position $(\vec{r})$ and time (t).

In most cases, we are interested in the states (eigenstates) with precise energy $(\mathcal{E})$. The wave function then has a separable form $\Psi(\vec{r}, t) = \psi(\vec{r})\phi(t)$, where $\psi(\vec{r})$ is a function of $\vec{r}$ only and $\phi(t)$ is a function of t only. $\phi(t)$ can be then removed from Eq. 2.12, hence the time-independent Schrödinger's equation in one dimension (1D) is obtained

$$\left[-\frac{\hbar^2}{2m}\frac{\partial^2}{\partial x^2} + V(x)\right]\psi(x) = \mathcal{E}\psi(x). \tag{2.13}$$

2.1.3.1 Dispersion relation

In general, boundary conditions are applied to Eq. 2.13 to obtain physical solutions. For an electron in a 1D space $(0 < x < L)$, there are two types of boundary conditions which can be imposed: (i) fixed boundary condition ($\phi(0) = 0$ and $\phi(L) = 0$) for the electron which is confined in this space, and (ii) periodic boundary condition ($\phi(0) = \phi(L)$ and $\frac{\partial\psi}{\partial x}(0) = \frac{\partial\psi}{\partial x}(L)$) for the traveling electron wave, for example, in solid crystals.

If $V(x) = V_0$, the solution of the Schrödinger's equation when applying the fixed condition for an electron is

$$\psi(x) = \sqrt{\frac{2}{L}}\sin(kx), \tag{2.14}$$

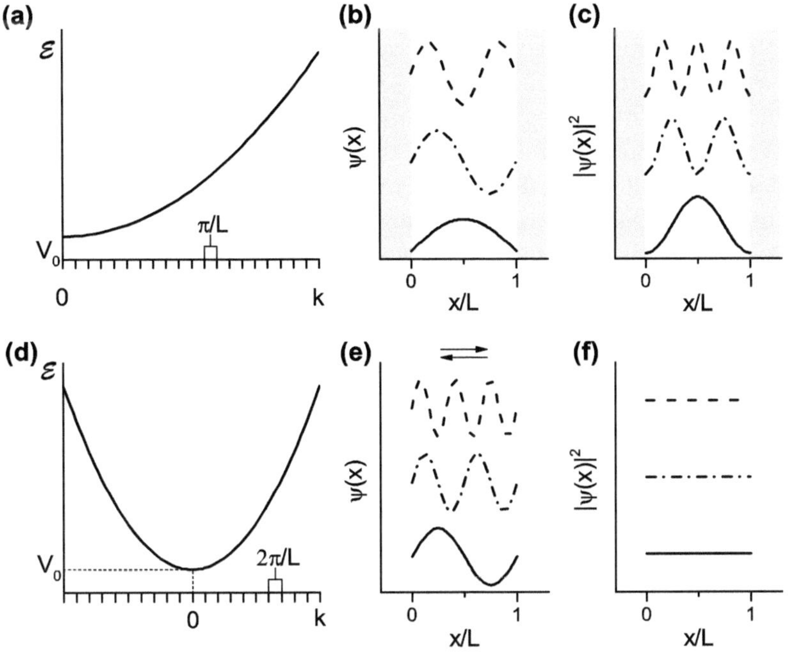

Figure 2.4 Dispersion relations of an electron in a 1D space when the (a) fixed and (d) periodic boundary conditions are applied. (b) Illustration of electron standing waves confined in the 1D space, corresponding to the dispersion relation in (a). Solid line: $k = \frac{\pi}{L}$, dash-dot line: $k = \frac{2\pi}{L}$, dashed line: $k = \frac{3\pi}{L}$. (c) Probability of finding the electron at a position x for the modes shown in (b). (e) Illustration of electron traveling waves, corresponding to the dispersion relation in (d). Solid line: $k = \frac{2\pi}{L}$, dash-dot line: $k = \frac{4\pi}{L}$, dashed line: $k = \frac{6\pi}{L}$. (f) Probability to find the electron at a position x for the modes shown in (e).

where $k = \frac{n\pi}{L}$, $n = 1, 2, \cdots$. k is the wavenumber which defines the momentum ($\hbar k$) and kinetic energy ($\frac{\hbar^2 k^2}{2m}$) of the electron. Hence, the energy of the electron is expressed as

$$\mathcal{E}(k) = \frac{\hbar^2 k^2}{2m} + V_0. \tag{2.15}$$

This is defined as the **dispersion relation** (or electronic structure) of the electron. It should be noted that k can only take certain values which are the multiples of $\frac{\pi}{L}$, as indicated in Fig. 2.4a. This shows that both the momentum and energy of the electron are quantized.

It is seen from Eq. 2.14 that standing waves are formed when the fixed boundary condition is applied, i.e., an electron is confined within the gray "walls" in Fig. 2.4b ($0 < x < L$). The solid, dash-dot, and dashed lines correspond to $n = 1, 2, 3$ for k, respectively. The probability to find the electron in position x is described by $|\psi(x)|^2$, as illustrated in Fig. 2.4c. It indicates that the probability of finding the electron is higher at certain locations. This agrees with the Heisenberg uncertainty principle, which suggests that the precise location of a quantum object is unknown if on the other hand the momentum has a precise value.

For electron transport in solid crystals, it is of interest to find the solution of the Schrödinger's equation when the periodic boundary condition is applied. Hence,

$$\psi(x) = \frac{1}{\sqrt{L}} \exp(ikx), \tag{2.16}$$

where $k = \frac{2n\pi}{L}$, $n = 0, \pm 1, \pm 2, \cdots$. It is worth noting that the dispersion relation has the same expression as that for the standing waves, i.e., Eq. 2.15, except that k has different values and it can be negative, as shown in Fig. 2.4d. The $\pm$ sign for n (or k) denotes the right/left direction the electron wave is traveling in, as demonstrated in Fig. 2.4e. In contrast to standing waves, the probability to find a traveling electron is the same at any location x, as illustrated in Fig. 2.4f.

2.1.3.2 Nanoelectronic structures

On the basis of whether a charge carrier is confined in certain direction(s), we are able to define certain structures such as a quantum dot, a quantum wire, and a quantum well.

If electrons are confined in all three dimensions in a structure, as illustrated in Fig. 2.5a, the wave function is described using Eq. 2.14. This means the electrons form standing waves in all directions, and there is no free momentum. This structure is then zero-dimensional (0D), and it is defined as a **quantum dot**. When one of the dimensions of the structure is long enough so that the fixed boundary condition for standing waves does not apply, electrons can then travel along this direction (where the dispersion

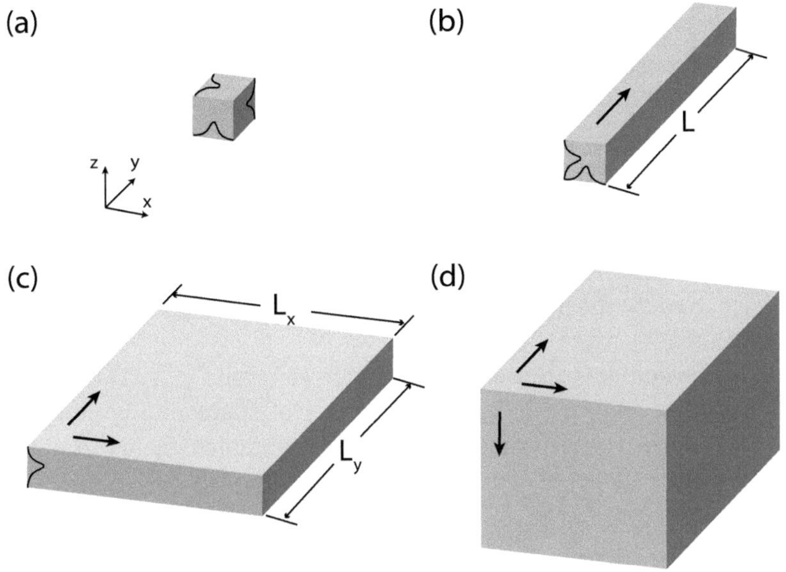

Figure 2.5 (a) 0D (quantum dot), (b) 1D (quantum wire), (c) 2D (quantum well), and (d) 3D structures commonly seen in nanoelectronic devices. The bell-shaped curves indicate that electrons are confined in the demonstrated dimension(s), and the arrows denote the direction(s) that electrons can move freely.

relation is described by Fig. 2.4d). This structure is therefore one-dimensional (1D), and it is sometimes defined as a **quantum wire**, as demonstrated in Fig. 2.5b. Similarly, a sheetlike two-dimensional (2D) structure, where electrons are only confined in one dimension, is called a **quantum well**, as shown in Fig. 2.5c. If electrons are able to travel freely in all three dimensions, the structure is three-dimensional (3D), as illustrated in Fig. 2.5c. Bulk material is a common example of a 3D conductor.[1]

[1]It is worth noting that for periodic potential (in section 2.2.1), we are interested in the properties of electrons near the conduction/valance band edges, not those whose momenta are close to zero. In this case, an electronic state is approximated by a sum over the Bloch functions and it can be described by an effective Schrödinger's equation using the effective mass theory. Readers who are interested in the details are recommended to consult [Davies (1998)]. The resulting concepts are the same as those in this subsection.

2.1.3.3 Density of states

The dispersion relation of an electron is important for the study of electrons in solid crystals, because (i) the group velocity can be determined by the derivative $v_g = \frac{\partial \mathcal{E}}{\hbar \partial k}$, and (ii) the density of states, $D(\mathcal{E})$, derived from this relation gives the number density of electrons (Eq. 2.10).

As defined in section 2.1.2, $D(\mathcal{E})$ represents the number of quantum states per energy, i.e., $D(\mathcal{E}) = \frac{\partial N_D}{\partial \mathcal{E}}$. A direct way to calculate N_D is to count the number of states in momentum space with a given value of k.

An example is to calculate $D(\mathcal{E})$ for electrons in a 1D quantum wire. As shown in Fig. 2.4d, the number of states in a momentum space with a given value of k can be obtained by dividing $2k$ (because k could be positive and negative) with the momentum quantum $\left(\frac{2\pi}{L}\right)$,

$$N_D = \frac{2k}{2\pi/L} = \frac{L\sqrt{2m(\mathcal{E} - V_0)}}{\pi\hbar}. \tag{2.17}$$

Note that Eq. 2.15 is used. Then the density of states for a 1D structure is

$$D(\mathcal{E}) = \frac{\partial N_D}{\partial \mathcal{E}} = \frac{L}{2\pi\hbar}\sqrt{\frac{2m}{\mathcal{E} - V_0}}. \tag{2.18}$$

For a 2D structure, the dispersion relation is the same as Eq. 2.15 except for $k^2 = k_x^2 + k_y^2$ here, where $k_x = \frac{2n\pi}{L_x}$ and $k_y = \frac{2n\pi}{L_y}$, as

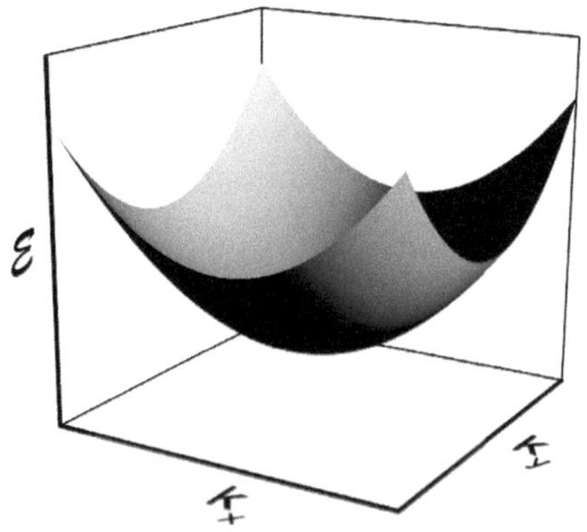

Figure 2.6 Dispersion relation for a 2D structure.

Table 2.1 Summary of number/density of quantum states for various structures

	1D	2D	3D
N_{D}	$g\dfrac{L\sqrt{2m(\mathcal{E}-V_0)}}{\pi\hbar}$	$g\dfrac{L_x L_y m}{2\pi\hbar^2}(\mathcal{E}-V_0)$	$g\dfrac{L_x L_y L_z}{6\pi^2\hbar^3}[2m(\mathcal{E}-V_0)]^{\frac{3}{2}}$
$D(\mathcal{E})$	$g\dfrac{L}{2\pi\hbar}\sqrt{\dfrac{2m}{\mathcal{E}-V_0}}$	$g\dfrac{L_x L_y m}{2\pi\hbar^2}$	$g\dfrac{L_x L_y L_z m}{2\pi^2\hbar^3}\sqrt{2m(\mathcal{E}-V_0)}$

illustrated in Fig. 2.6. The subscripts x and y denote the dimensions, as shown in Fig. 2.5. Thus the expressions for the number, N_{D}, and density, $D(\mathcal{E})$, of quantum states can be derived.

Table 2.1 summarizes the calculated N_{D} and $D(\mathcal{E})$ for 1D, 2D, and 3D structures. Note that g is the degeneracy of electrons which is caused by a variety of reasons (spin, valley, etc.). The derivation processes for N_{D} and $D(\mathcal{E})$ in 2D and 3D structures are straightforward and thus not shown here.

Figure 2.7 shows the calculated density of states for 0D, 1D, 2D, and 3D structures using the equations in Table 2.1. Note that $D(\mathcal{E})$ for a 0D structure is expressed by a δ function. This information is particularly useful for calculating the charge carrier density in solid materials.

It is interesting noted that the number of quantum states can also be calculated using statistical mechanics. As shown in Fig. 2.2 and Eq. 2.8, N_{D} for a 3D structure is obtained,

$$\begin{aligned} N_{\mathrm{D}} &= g\int\frac{\mathrm{d}p_1\mathrm{d}p_2\cdots\mathrm{d}p_r\mathrm{d}x_1\mathrm{d}x_2\cdots\mathrm{d}x_r}{h^r} \\ &\overset{r=3}{=} g\int 4\pi\hbar^3 k^2\frac{\mathrm{d}k\mathrm{d}x_1\mathrm{d}x_2\mathrm{d}x_3}{h^3} \\ &= g\int\frac{2\pi L_x L_y L_z}{h^3}(2m)^{\frac{3}{2}}\mathcal{E}^{\frac{1}{2}}\mathrm{d}\mathcal{E}. \end{aligned} \tag{2.19}$$

It can be seen that this equation is exactly the same as that in Table 2.1. Readers are encouraged to derive the expressions of N_{D} for 1D and 2D structures using this method.

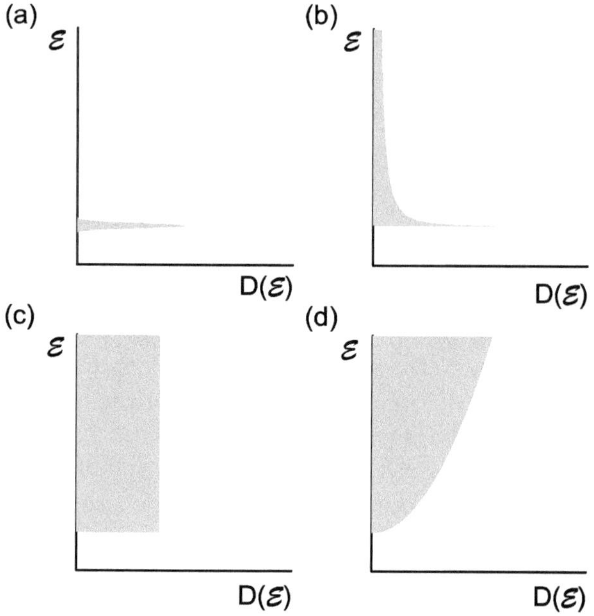

Figure 2.7 Density of states for (a) 0D (quantum dot), (b) 1D (quantum wire), (c) 2D (quantum well), and (d) 3D structures.

2.2 Nearly Free Electron Gas

Modern condensed matter physics investigates the physical processes (such as electron transport) in crystalline solids, where the atoms are arranged in periodic arrays. These structures can be found in everyday minerals, such as rock salt, diamond, quartz, among others. In 1913, W. Bragg provided an experimental evidence for the crystallinity using X-ray diffraction. This discovery not only opened the field of crystallography, but also had a great impact on the theory of charge transport phenomena (dispersion relation) in solids.

2.2.1 Modification to Dispersion Relation

Valence electrons (as opposed to core electrons at the atomic level) in crystals are of importance for solid state physics, as they are often the majority charge carriers.

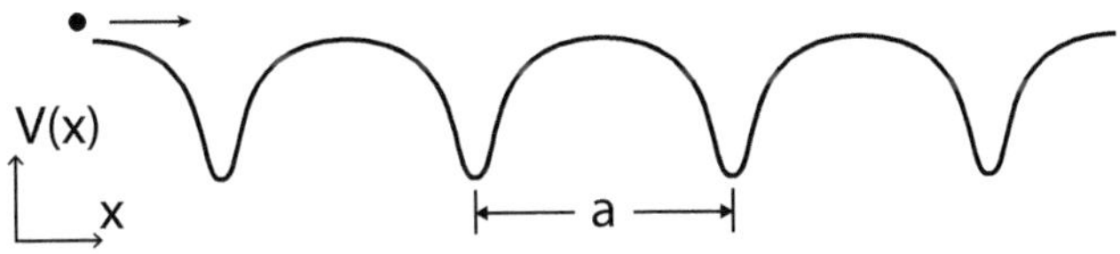

Figure 2.8 An electron (black sphere[2]) traveling in a positive x direction within a periodic potential profile, where a represents the crystal period.

When these electrons are traveling in the crystal lattice, they experience the effective potential from the positive ions and the core electrons. Since the atoms are orderly arranged in the lattice, the potential profile is also periodic, as illustrated in Fig. 2.8. Therefore, solutions different to those found in section 2.1.3.1 can be obtained, as $V(x)$ is no longer a constant.

It can be seen from Fig. 2.8 that $V(x)$ is dependent on the interatomic distance, a, in crystals. As atoms can be arranged in different ways to form various structures, it is of importance to review the common crystal lattices.

2.2.1.1 Crystal structure

Ideally, a crystal is constructed by atom groups infinitely repeating in certain directions. Such an atom group is a **basis**, and the geometric pattern where the basis is attached is **lattice**. Since graphene is a 2D material, this section focuses on the discussion of 2D structures. Readers interested in 3D crystal structures are recommended to consult [Ashcroft and Mermin (1976); Kittel (2004)].

Figure 2.9 illustrates a 2D crystal, where a basis consisting of two atoms is highlighted in a dash-dotted ellipse. The dashed oblique grid represents the lattice. Note that the gray parallelogram is a **primitive cell** of the structure. A primitive cell is defined as a unit cell of the lattice, which can fill the whole space of the crystal structure through repetitive translation operations. It can be chosen in many different ways for a lattice. However, the number of atoms primitive cells of different choices (or basis) is always identical for a given crystal.

[2] In the view of quantum mechanics, the electron is considered as a wave. The physical image of electron wave traveling in periodic potential can be found in [Kittel (2004)].

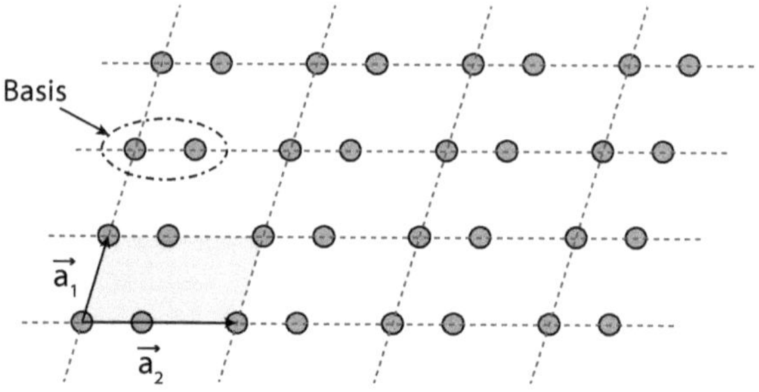

Figure 2.9 Illustration of a 2D crystal structure, where a basis highlighted in the dash-dotted ellipse and a primitive cell is indicated by the gray parallelogram. The dashed oblique grid represents the lattice.

The primitive cell is geometrically defined by two **primitive translation vectors** $\vec{a}_1$ and $\vec{a}_2$, as shown in Fig. 2.9. It is seen that the position of any node ($\vec{r}_i$) in the lattice (dashed grid) can be expressed by the integer multiples of $\vec{a}_i$:

$$\vec{r}_i = u_i\vec{a}_1 + v_i\vec{a}_2, \quad (2.20)$$

where u_i and v_i are arbitrary integers. This suggests that the lattice is generated from integer translations (Eq. 2.20) of the primitive cell. These vectors are important as they provide indications for the periodic potentials in Fig. 2.8.

It also can be seen from Fig. 2.9 that crystal lattices can be symmetric, i.e., lattices can be mapped onto themselves using certain operations (often rotations). As $\vec{a}_1$ and $\vec{a}_2$ are arbitrarily chosen, if we want to ensure that the lattice is symmetric, some restrictive conditions must be imposed on these vectors. These restrictions lead to **special lattice types (Bravais lattices)**. There are five 2D Bravais lattices, including square, rectangular, centered rectangular, hexagonal, and oblique lattices [Kittel (2004)].

2.2.1.2 Reciprocal lattice and bandgap

Similar to the idea of electron transport in section 2.1.3, each valley in the potential profile shown in Fig. 2.8 can be thought of a quantum "well" for traveling electrons. The solutions to the Schrödinger's equation (Eq. 2.13) involving periodic potential ($V(x) = V(x + a)$)

can be expressed as

$$\psi(x) = \exp(ikx) \cdot u(x), \tag{2.21}$$

where $u(x) = u(k + a)$ is a periodic function. Equivalently, this equation can also be written as

$$\psi(x + a) = \exp(ika) \cdot \psi(x). \tag{2.22}$$

This is **Bloch's theorem**. Note that the period of the wave function $\psi(x)$ is not necessarily a, unless $k = \frac{2n\pi}{a}$ where n is an arbitrary integer.

The periodic potential $V(x)$ can be written as a Fourier series

$$V(x) = \sum_{n=-\infty}^{+\infty} V_n \exp(iG_n x), \tag{2.23}$$

where V_n is a constant, and $G_n = \frac{2n\pi}{a}$ $(n = 0, \pm 1, \pm 2, \cdots)$ is the **reciprocal lattice vector**. It is then straightforward to see that $\exp(iG_n a) = 1$, and hence the period for $\exp(\pm iG_n x)$ is also a, because $\exp(\pm iG_n x) = \exp(\pm iG_n x) \cdot \exp(iG_n a) = \exp[iG_n(\pm x + a)]$.

If we redefine $u(x)$ in Eq. 2.21 as $u'(x) = \exp(-iG_n x) \cdot u(x)$, the period of $u'(x)$ is still a. Eq. 2.21 can be re-written as

$$\begin{aligned} \psi(x) &= \exp[i(k + G_n)x] \cdot \exp(-iG_n x) \cdot u(x) \\ &= \exp(ik'x) \cdot u'(x), \end{aligned} \tag{2.24}$$

where

$$k' = k + G_n. \tag{2.25}$$

This is a very important relation indicating that *we can relabel the same quantum state with a different crystal momentum k'*, rather than k only.

It suggests that the energy of a traveling electron in crystals can be represented by more equivalent states which are indistinguishable from the original state (Fig. 2.4d). These equivalent representations are produced by repeating the original dispersion relation by G_n along the k axis, as illustrated in Fig. 2.10a. It can be seen that the parabolic dispersion relations cross each other when $k = \frac{G_n}{2}$, these locations are particularly important in solid state physics, and they define the Brillouin zones. Figure 2.10b highlights the energy band structure for a 1D lattice within the first **Brillouin**

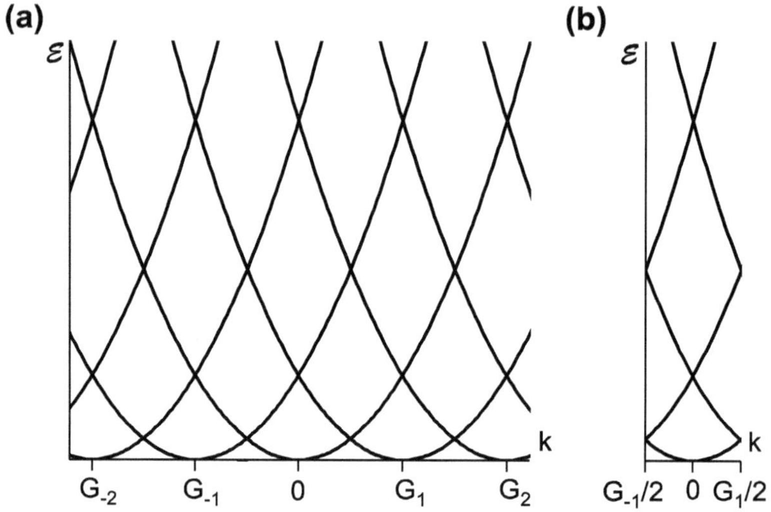

Figure 2.10 Energy versus wave vector for an electron traveling in crystals with small amplitude of periodic potentials in (a) full k space, and (b) the first Brillouin zone $\left(\frac{G_{-1}}{2} < k < \frac{G_1}{2}\right)$.

zone ($\frac{G_{-1}}{2} < k < \frac{G_1}{2}$). Note that this figure is an approximation of the precise band structure when the amplitude of the periodic potential ($V(x)$) is very small.

In general, the amplitude of the periodic potential, $V(x)$, is not negligible. To calculate the energy band structure more accurately, we need to substitute Eq. 2.23 into Eq. 2.13. There are several methods to solve the equation, including the Kronig–Penney model [Kittel (2004)], perturbation theory [Szkopek (2013)], direct application of Bloch's theorem, and so on. The results suggest that the dispersion relation is mostly parabolic, except near the cross points ($\frac{G_n}{2}$) in Fig. 2.10 which can be expressed as

$$\mathcal{E} = \frac{\left(\hbar\frac{G_n}{2}\right)^2}{2m} + V_0 \pm |V_n| + \frac{(\hbar q_n)^2}{2m^*}, \tag{2.26}$$

where $q_n = k - \frac{G_n}{2}$, and $\frac{1}{m^*} = \frac{1}{m}\left[1 \pm \frac{(\hbar G_n)^2/2m}{2|V_n|}\right]$ defines the inverse of **effective mass** of the electron.

Figure 2.11 shows the energy band structure in a 1D lattice using Eq. 2.26, which becomes discontinuous when $k = \frac{G_n}{2}$. This is

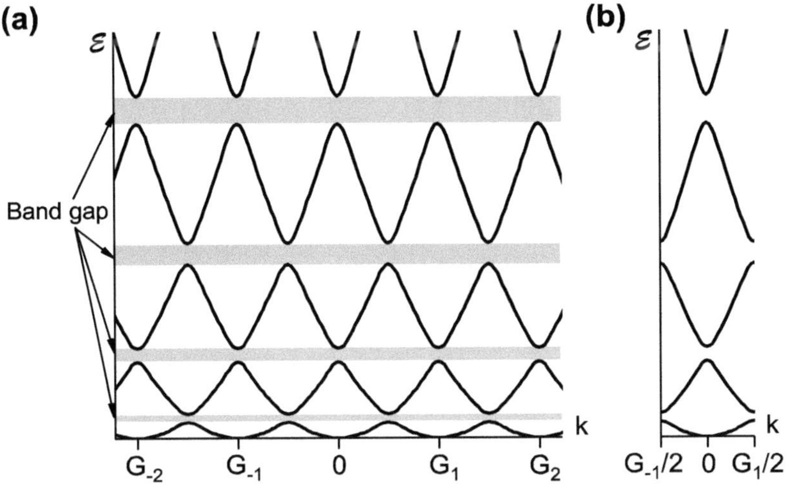

Figure 2.11 Energy band structure of a 1D lattice plotted in (a) the periodic, and (b) the reduced (first) Brillouin zone schemes.

the direct result of the interaction between traveling and scattered electron waves due to the presence of the non-negligible periodic potential. These discontinuities are defined as **bandgaps** ($\mathcal{E}_{\mathrm{g}} = 2|V_n|$, indicated in Fig. 2.11), and they are extremely important in understanding the electrical conductivity of solid crystals. As the energy band structure repeats itself in the periodic zone scheme (with the period of G_1), we are often interested in the first Brillouin zone shown in Fig. 2.11b.

Figure 2.12a shows a magnification of the band structure near the first Brillouin zone edge ($\frac{G_{-1}}{2}$), where the **conduction** and the **valence** bands are defined by convention. Note that $q = k - \frac{G_{-1}}{2}$ is used as the horizontal axis in Fig. 2.12a, instead of k. It can be seen from Eq. 2.26 and Fig. 2.12a that the energy vs. q relation near the bandgap is also parabolic. This is similar to the dispersion relation in section 2.1.3.1 where the potential is constant. It then suggests that an approximation can be made such that the periodic potential profile is equivalent to a familiar quantum well, as shown in Figs. 2.12b–c. The precise solution to the Schrödinger's equation for the potential in Fig. 2.12b is illustrated in (d), where the gray dashed curve is the slow varying envelope. Note that this envelope is the

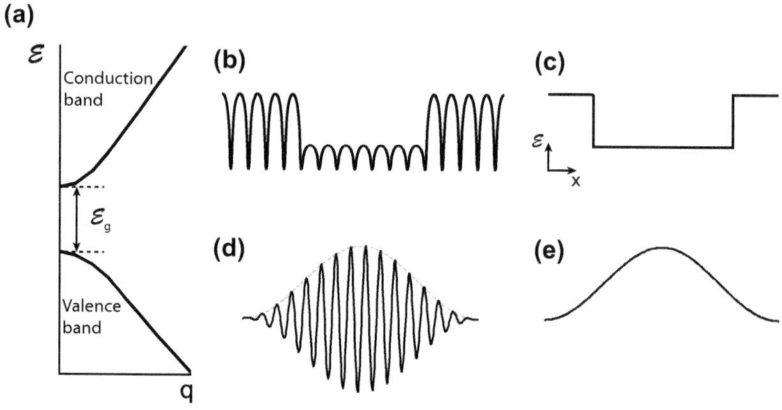

Figure 2.12 (a) Energy vs. q relation near a bandgap ($\mathcal{E}_g$), where the conduction and the valence bands are labeled. (b) Periodic potential for a 1D quantum well. (c) An equivalent quantum well of (b) constructed by the conduction band edge. (d) An electron wave function in the quantum well in (b), where the gray dashed curve is the slow varying envelope. (e) The solution of the effective Schrödinger's equation using the potential profile in (c).

same as the wave function in Fig. 2.12e, which satisfies an effective Schrödinger's equation for the quantum well shown in (c).

This model uses the effective mass of electrons, and hence it is defined as the **effective mass theory**. It does not need to describe the wave functions at the atomic level, and hence simplifies the investigation of electrons near the bandgaps. Moreover, the nanoelectronic structures described in section 2.1.3.2 are still useful.

For a 2D or 3D crystal structure, the reciprocal vectors ($\vec{G}$) construct a reciprocal lattice space, i.e., $\vec{G} = \sum_i u_i \vec{b}_i$ where u_i is an arbitrary integer, and $\vec{b}_i$ is a primitive vector. In general, the primitive reciprocal lattice vector is defined as

$$\vec{b}_i \cdot \vec{a}_j = 2\pi \delta_{ij}, \tag{2.27}$$

where $\delta_{ij} = 1$ if $i = j$ and $\delta_{ij} = 0$ if $i \neq j$. For a 2D lattice,

$$\vec{b}_1 = 2\pi \frac{\vec{a}_2 \times (\vec{a}_1 \times \vec{a}_2)}{\vec{a}_1 \cdot [\vec{a}_2 \times (\vec{a}_1 \times \vec{a}_2)]}, \tag{2.28a}$$

$$\vec{b}_2 = 2\pi \frac{\vec{a}_1 \times (\vec{a}_2 \times \vec{a}_1)}{\vec{a}_2 \cdot [\vec{a}_1 \times (\vec{a}_2 \times \vec{a}_1)]}. \tag{2.28b}$$

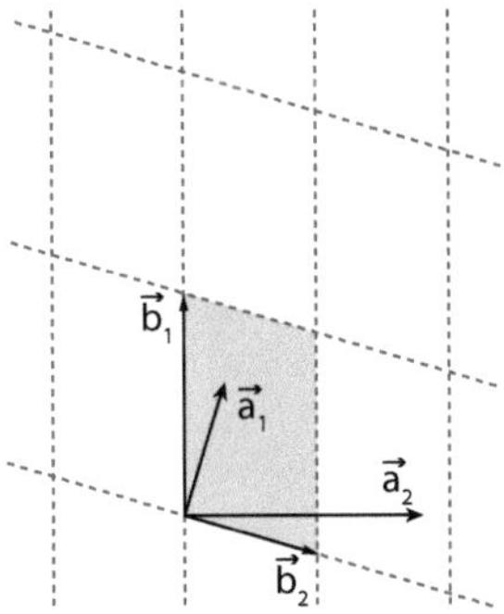

Figure 2.13 The reciprocal lattice for the 2D crystal shown in Fig. 2.9.

Figure 2.13 illustrates the reciprocal lattice of the 2D crystal shown in Fig. 2.9, which is useful for calculating the band structure.

2.2.2 Density of States

As we are often interested in the electrical properties near the bandgap, it then is important to calculate the density of states. Since the effective mass theory suggests that the dispersion relation is still parabolic close to the bandgaps, the results for the density of states in section 2.1.3.3 are still valid, except the effective mass (Eq. 2.26) should be used instead.

Figure 2.14a shows the density of states near a bandgap for a 2D crystal structure. This is similar to Fig. 2.7c except that the conduction band (CB) and the valence band (VB) are both included.

In Fig. 2.14b an example of quantum well is shown, where a 2D GaAs sheet (the well) is encapsulated by two AlGaAs layers (the barriers). Although the strict derivation process for solving the Schrödinger's equation is complicated, the idea is similar to the discussion in section 2.1.3.1, where n can take different numbers for k in Eq. 2.14. This allows the formation of various modes (**sub-bands**) for the standing electron wave in the z direction, which are shown in Fig. 2.14c. These sub-bands lead to the additional density of states as illustrated in Fig. 2.14d, as commonly seen in literature.

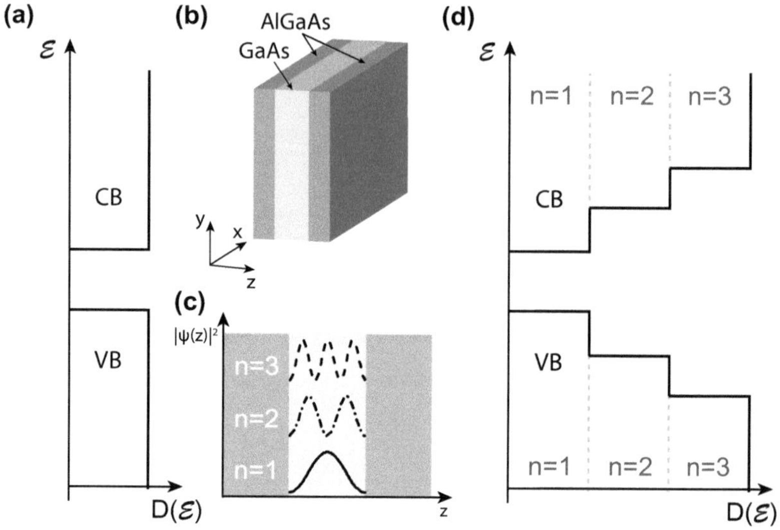

Figure 2.14 (a) Density of states near a bandgap of a 2D lattice, where the conduction band (CB) and the valence band (VB) are shown. (b) An AlGaAs/GaAs/AlGaAs quantum well. (c) The first three allowed modes (sub-bands) of an electron wave in the quantum well shown in (b). (d) Density of states near the bandgap for the structure shown in (b), including the contribution from all three sub-bands.

2.2.3 Electronic Properties of Semiconductors

The Drude model (Eq. 2.7) suggests that the mobility (μ, discussed in section 2.1.1.1) and the charge carrier density (n) determines the electrical behavior of a semiconductor. In previous subsections, efforts have been made to discuss the density of states, $D(\mathcal{E})$, for nanoelectronic structures. This is because $D(\mathcal{E})$ plays an important role when studying the charge carrier density in semiconductors, as shown in Eq. 2.10. Moreover, when the size of a semiconductor device is greatly reduced (typically less than 10 nm), quantum capacitance (a phenomenon related to $D(\mathcal{E})$) becomes important.

2.2.3.1 Charge carrier density

For electrons in the conduction band of a 2D lattice (where the band edge is taken as 0 for simplicity), it is straightforward to derive the expression for the charge carrier density from Eq. 2.10,

$$\begin{aligned} n_{2D} &= \frac{1}{L_x L_y}\int_0^{\infty} f_{FD}(\mathcal{E})D_{2D}(\mathcal{E})d\mathcal{E} \\ &= \frac{m^* k_B T}{\pi\hbar^2}\ln\left[1+\exp\left(\frac{\mathcal{E}_F}{k_B T}\right)\right], \end{aligned} \tag{2.29}$$

where $\mathcal{E}_F$ is the Fermi level defined in section 2.1.2, and m^* is the effective mass shown in Eq. 2.26.

Similarly, for electrons in the conduction band of a 3D lattice,

$$n_{3D} = N \cdot F_{1/2}(\eta), \tag{2.30}$$

where $N = 2\left(\frac{m^* k_B T}{2\pi^2\hbar^2}\right)^{\frac{3}{2}}$, $\eta = \frac{\mathcal{E}_F}{k_B T}$, and

$$F_j(\eta) = \frac{1}{\Gamma(j+1)}\int_0^{\infty}\frac{\xi^j}{\exp(\xi-\eta)+1}d\xi \tag{2.31}$$

is the **Fermi–Dirac integral** of order j. $\xi = \frac{\mathcal{E}}{k_B T}$, and Γ is the Gamma function.

Figure 2.15 illustrates a Fermi–Dirac distribution aligned with the density of states for a 2D and 3D structure. This indicates that a **degenerate electron gas** is formed in these structures, because the electrons are degenerate (follow Fermi–Dirac distribution) when the Fermi level ($\mathcal{E}_F$) is above the conduction band (CB) minimum.

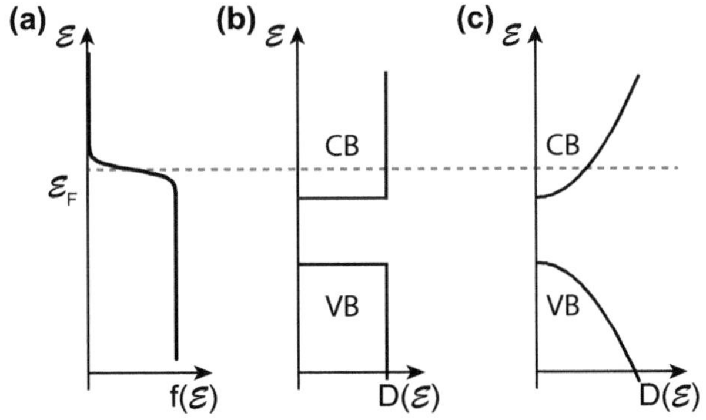

Figure 2.15 (a) Fermi–Dirac distribution aligned with the density of states for (b) a 2D and (c) a 3D crystal structure, where electrons are degenerate in this case.

Similarly, a degenerate hole gas is formed when the Fermi level is below the valence band (VB) maximum.

As suggested in section 2.1.2, when $\mathcal{E} \gg \mu_p$ (or $\mathcal{E} \gg \mathcal{E}_F + k_B T$), the Fermi–Dirac statics approximates to the Maxwell–Boltzmann distribution, i.e., $f(\mathcal{E}) \propto \exp\left(-\frac{\mathcal{E}}{k_B T}\right)$. A classical **non-degenerate electron gas** is then formed. Hence, the charge carrier densities in 2D and 3D structures become

$$n_{2D} \approx \frac{m^* k_B T}{\pi \hbar^2} \exp\left(\frac{\mathcal{E}_F}{k_B T}\right), \tag{2.32}$$

and

$$n_{3D} \approx N \exp\left(\frac{\mathcal{E}_F}{k_B T}\right), \tag{2.33}$$

Note that $\mathcal{E}_F$ is slightly below the band edges (0) in this case. It is seen that the band edges are important in evaluating the charge carrier density, and they are included in commonly seen band diagrams (see section 2.3).

2.2.3.2 Quantum capacitance

Capacitor is one of the most important electrical elements in electronic devices. **Quantum capacitance** (C_q), an intrinsic property, of capacitors made of semiconductors plays a critical role when the size of the capacitors reduces into nanoscales. This concept is introduced in semiconductor physics while studying transistors (as the gate capacitance defines the number of charge carriers in the conducting channel) [Luryi (1988); Szkopek (2013)].

Unlike inversion layer capacitance (see section 2.3) which is determined by the thickness of the inversion layer at the oxide-semiconductor interface, quantum capacitance is limited by the Pauli exclusion principle. This is because the charge carrier density (n) in a semiconductor is defined by the integration of the density of states, $D(\mathcal{E})$, of the material and the Fermi–Dirac distribution. Due to the relatively low $D(\mathcal{E})$ in semiconductors, this then provides a certain value for n (n^{lim}) at a given energy (Fermi level, $\mathcal{E}_F$). When $n = n^{\text{lim}}$, no more charge carriers can be added into the capacitor, unless extra energy is added (i.e., increase $\mathcal{E}_F$).

Figure 2.16a,b illustratively describes this process. A capacitor in (a) is able to store more charges when n is below the limit n^{lim}.

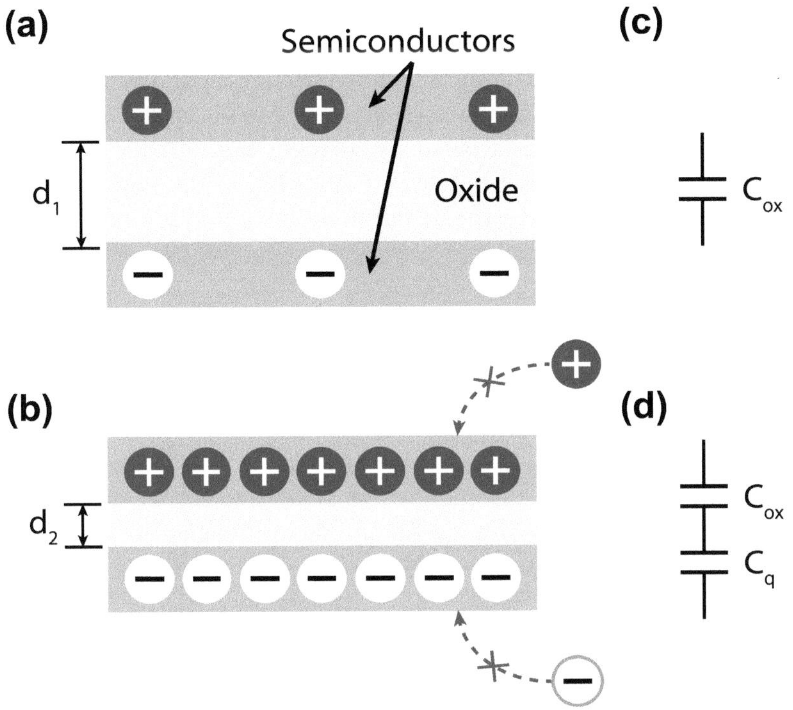

Figure 2.16 (a, b) Illustration of parallel plate capacitors with different oxide layer thicknesses ($d_1 > d_2$). (c, d) Equivalent circuit model for the capacitors shown in (a) and (b), respectively.

The quantum capacitance does not take effect in this case. Hence, the equivalent circuit element is simply a parallel plate capacitor as shown in (c). If the thickness of the oxide layer decreases, n becomes larger as the capacitance is inversely related to distance

$$C_{\text{ox}} = \frac{\epsilon}{d}, \tag{2.34}$$

where C_{ox} is the capacitance per unit area, ϵ is the permittivity of the oxide layer, and d is the oxide layer thickness. When $n = n^{\text{lim}}$, no more charge carriers can be added onto the semiconductive plates unless more energy is added to the charge carriers (i.e., increase the voltage applied on the capacitor), as shown in (b). The equivalent circuit is then C_{ox} and C_{q} in series, as depicted in (d). In this case,

C_q dominates the circuit, as C_{ox} is a very large number and smaller values are of greater significance.

Capacitance, in general, is defined as the capability of storing charges by a capacitor. Therefore, according to the earlier definition, quantum capacitance is given by dividing the number of stored charges by the applied potential (equivalently, the energy of the charge carriers). Hence, the quantum capacitance per unit area can be written as

$$C_{\mathrm{q}} = \frac{e\mathrm{d}n}{\mathrm{d}V} = e^2 \left.\frac{\mathrm{d}n}{\mathrm{d}\mathcal{E}}\right|_{\mathcal{E}=\mathcal{E}_{\mathrm{F}}} = e^2 D(\mathcal{E}_{\mathrm{F}}), \tag{2.35}$$

where $D(\mathcal{E}_{\mathrm{F}}) = \left.\frac{\mathrm{d}n}{\mathrm{d}\mathcal{E}}\right|_{\mathcal{E}=\mathcal{E}_{\mathrm{F}}}$ represents the density of states at the Fermi level.

The quantum confinement energy of an electron is around $\frac{1}{D(\mathcal{E}_{\mathrm{F}})}$, as $D(\mathcal{E}_{\mathrm{F}})$ represents the number of states (electrons, when $\mathcal{E} = \mathcal{E}_{\mathrm{F}}$) per energy. Therefore, the quantum confinement contribution to the electron addition energy can be represented by quantum capacitance, $\mathcal{E}_{\mathrm{q}} = \frac{e^2}{2C_{\mathrm{q}}} = \frac{1}{2D(\mathcal{E}_{\mathrm{F}})}$ [Szkopek (2013)]. Note this expression has the same form as the electrostatic charging energy $\mathcal{E}_{\mathrm{C}} = \frac{e^2}{2C}$.

2.3 Electrons in Heterostructures

For electrical applications, we often construct **heterostructures** using various solid materials. Electron energy can vary drastically at different positions within the heterostructures. It is then common to draw **band diagrams** where the conduction (CB) and valence band (VB) edges are plotted as a function of the position. This is because band edges are important in evaluating the charge carrier distribution, as discussed in section 2.2.3.1. Hence, band diagrams provide a direct physical image to study charge carriers in heterostructures.

2.3.1 Metals, Insulators, and Semiconductors

Figure 2.17a shows the energy band diagram for a solid crystal. If the Fermi level ($\mathcal{E}_{\mathrm{F}}$) is located inside the conduction band of a material,

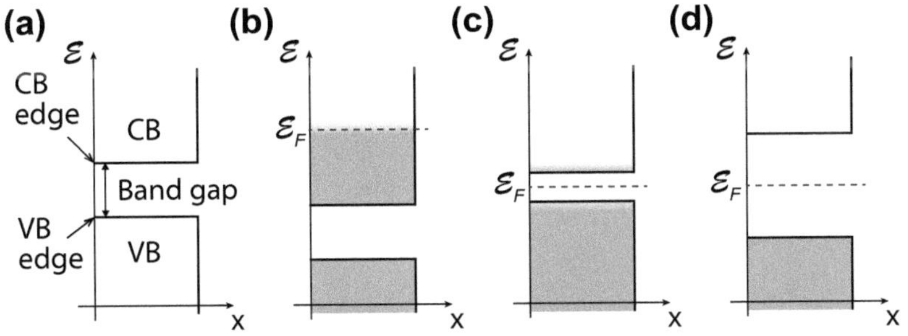

Figure 2.17 Energy band diagrams for (a) a solid crystal where the bandgap and the conduction and valence bands are indicated, (b) a metal, (c) a semiconductor, and (d) an insulator.

and it is far from the conduction band edge, the material is metal, as illustrated in Fig. 2.17b. For a semiconductor, the bandgap is small, and $\mathcal{E}_F$ generally sits within the bandgap, as shown in Fig. 2.17c. As the Fermi–Dirac distribution has a high energy tail (Fig. 2.3), there are intrinsic charge carriers in the conduction and valence bands due to thermal excitation. Figure 2.17d shows the band diagram of an insulator, where the bandgap is significantly wider than that of a semiconductor. In this case, there are no intrinsic charge carriers because of the large bandgap.

It can be concluded from Fig. 2.17c,d that the intrinsic charge carrier density in a semiconductor greatly depends on the bandgap, and the low carrier density of an intrinsic semiconductor is not enough for most applications. A common technique to increase the charge carrier density in semiconductors is to introduce dopants (different elements). These elements create additional $D(\mathcal{E})$ within the bandgap, and hence shift $\mathcal{E}_F$.

Common non-compound semiconductors are often made of group IV elements, such as silicon or carbon. When such a single crystalline semiconductor contains group V elements (e.g., nitrogen, phosphate), it is called n-doped. This is because the group V element contains one more electron, which can contribute to electrical conduction in the semiconductor. As an electron is negatively charged, this semiconductor is defined as **n-doped**.

Figure 2.18a shows an n-doped semiconductor with a moderate dopant concentration. As a result of the doping, $\mathcal{E}_F$ is moved to a

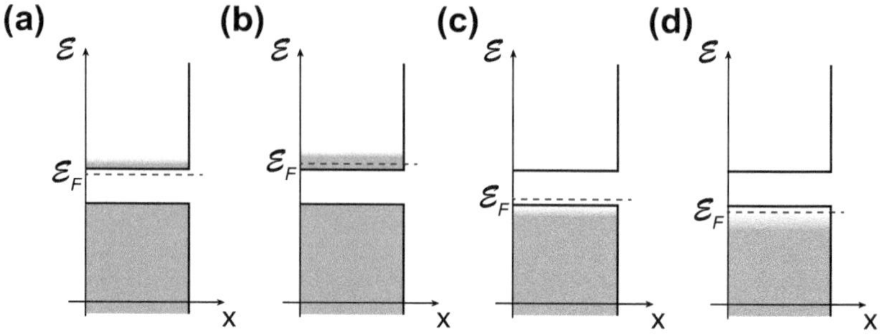

Figure 2.18 Energy band diagrams for (a) a moderately n-doped, (b) a heavily n-doped, (c) a moderately p-doped, and (d) a heavily p-doped semiconductor.

position close to the conduction band edge. This induces a higher electron density in the conduction band. $\mathcal{E}_F$ can be further shifted into the conduction band when the semiconductor is heavily n-doped, as illustrated in Fig. 2.18b. The electron density in this case is much higher, and the band diagram is similar to that shown in Fig. 2.17b for a metal. Therefore, the heavily doped semiconductor is said to be metallic on some occasions.

When electrons are excited from the valence band into the conduction band, **holes** are left in the valence band and they can contribute to electrical conductance [Kittel (2004)]. If holes are the majority charge carriers in a semiconductor (which usually contains group III elements, e.g., boron, gallium), it is **p-doped**. Figure 2.18c,d shows a moderately and a heavily p-doped semiconductor, respectively.

2.3.2 Heterostructures

Semiconductors with different doping levels, as well as metals and insulators, are often used to construct electrical devices (such as pn junctions and transistors) for different applications. It is then useful to draw the band diagrams for the heterostructures.

At equilibrium, the Fermi level of electron gas in different materials within a heterostructure is the same. This results in a redistribution of charge carriers at the interface of the materials in the heterostructure, which also changes the potential profile (incurs band bending) and affects the band diagram. To obtain the exact band diagram for a heterostructure, one needs to solve a set of equations including Schrödinger's equation, the Fermi–Dirac distribution, and Poisson's equation. However, on most occasions, an illustrative band diagram is enough to explain the physics. It is then common to use the technique of band diagram alignment to study heterostructures.

Band alignment for heterostructure is generally a four-step process [Szkopek (2013)]. Step 1: Align the vacuum level for the electrons in heterostructure due to Anderson's rule [Davies (1998)], and assume all the bandgap in each material is the same. This gives the flat-band voltage. Step 2: Align the Fermi levels, as the energy of electrons in an electrically neutral heterostructure is the same everywhere. Step 3: Connect the band edges. Step 4: Restore the difference in bandgaps, and take into account the potential perturbation by the charge carriers in accumulation or inversion layer if necessary.

In this section, we apply this technique to two popular structures, metal–oxide–semiconductor (MOS) and metal–semiconductor systems, as they are useful for the investigation of semiconductor electrochemistry, e.g., graphene–electrolyte systems.

2.3.2.1 Metal–oxide–semiconductor systems

Before alignment, the conduction and valance band edges for an oxide layer ($\mathcal{E}_{\mathrm{C,O}}$ and $\mathcal{E}_{\mathrm{V,O}}$) and a p-type semiconductor ($\mathcal{E}_{\mathrm{C,S}}$ and $\mathcal{E}_{\mathrm{V,S}}$), as well as the Fermi level for a metal ($\mathcal{E}_{\mathrm{F,M}}$) and the semiconductor ($\mathcal{E}_{\mathrm{F,S}}$), are shown in Fig. 2.19a. Note that this band diagram can only be obtained when applying a negative bias on the metal ($V_{\mathrm{g}} = V_{\mathrm{fb}}$, where V_{g} is the voltage between metal and semiconductor, and V_{fb} is the flat-band voltage), as $\mathcal{E}_{\mathrm{F,M}}$ is higher than $\mathcal{E}_{\mathrm{F,S}}$ when the vacuum level of all materials are aligned. Hence, this condition is defined as the **flat-band condition**.

When V_{fb} is removed ($V_{\mathrm{g}} = 0$), $\mathcal{E}_{\mathrm{F,M}}$ is aligned with $\mathcal{E}_{\mathrm{F,S}}$ as discussed previously. This results in, as shown in Fig. 2.19b, $0 = V_{\mathrm{fb}} + \varphi_{\mathrm{ox}} + \varphi_{\mathrm{s}}$, where φ_{ox} and φ_{s} represents the potential drops within the oxide layer and at the oxide–semiconductor interface. φ_{s} becomes non-zero because of the diffusion of charge carriers at the

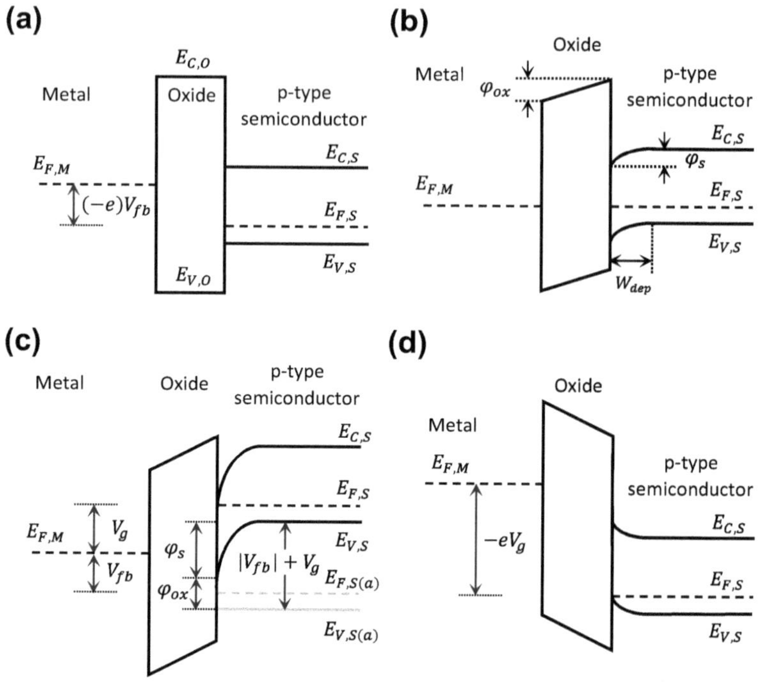

Figure 2.19 Band diagram schematics of (a) flat-band condition, (b) depletion layer with zero bias, (c) inversion layer, and (d) accumulation layer.

oxide–semiconductor interface due to the difference in Fermi levels in the heterostructure. The concentration of charge carriers at the oxide–semiconductor interface decreases in this case, because $\mathcal{E}_{F,S}$ is further away from both $\mathcal{E}_{C,S}$ and $\mathcal{E}_{V,S}$. Therefore, a **depletion layer** is formed at the interface.

When the gate voltage becomes positive on the metal, relative to the semiconductor, $\mathcal{E}_{F,M}$ gets lower compared to $\mathcal{E}_{F,S}$, as shown in Fig. 2.19c. This introduces more electrons at the interface of the oxide and the semiconductor, as $\mathcal{E}_{F,S}$ is close to $\mathcal{E}_{C,S}$. This layer of electron gas formed at the interface is defined as the **inversion layer**, because the semiconductor is p-type.

As $V_{fb} = \frac{\mathcal{E}_{F,M}-\mathcal{E}_{F,S}}{-e}$ in Fig. 2.19a, V_{fb} is a negative value. Since $V_g = \frac{\mathcal{E}_{F,M}-\mathcal{E}_{F,S}}{-e}$ in Fig. 2.19c, V_g is then positive. A useful relation between V_g

and the potential drops at the oxide–semiconductor interface and within the oxide layer can be written as

$$V_g = V_{fb} + \varphi_{ox} + \varphi_s. \tag{2.36}$$

This relation is not only important for the MOS system, but also for many other heterostructures as will be discussed later. Figure 2.19d indicates a negative V_g on metal, where an **accumulation layer** forms.

In summary, Fig. 2.19 shows the band diagrams of (a) flat-band condition, (b) formation of a depletion layer with zero bias, (c) an inversion layer with positive bias (on the metal), and (d) an accumulation layer with negative bias at the oxide–semiconductor interface in a MOS structure. It should be noted that the carriers in the surface depletion layer of semiconductor also follow Boltzmann statistics, as $\mathcal{E}_{F,S}$ is away from the band edges at the oxide–semiconductor interface. In contrast, degenerate electron/hole gas is formed in inversion/accumulation layer since the bent band edges are close to the $\mathcal{E}_{F,S}$.

2.3.2.2 Metal–semiconductor systems

In contrast to semiconductors, there is no band edge in metals in energy band diagrams, as seen in last section. Since the intrinsic charge carrier density in metal is higher than that in semiconductor and electrolyte, the screening length in metal is lower. This affects the energy band bending at interfaces involving metals. One of the aims of the book is to understand the band alignment process of heterostructures where metals are in contact with electrolyte. It is then useful to first investigate the well-studied metal–semiconductor junction in semiconductor physics.

Figure 2.20 describes the band diagrams before (a, c, e, g) and after (b, d, f, h) the metal contacts the semiconductor, where (a–d) and (e–h) are for p-type and n-type semiconductors, respectively. By Anderson's rule [Davies (1998); Szkopek (2013)], the vacuum levels ($\mathcal{E}_{vac}$) of different phases are aligned before contact to get the flat-band voltage V_{fb}. After contact, the electrons with higher energy (greater value of $\mathcal{E}_F$) in a phase intend to diffuse or transfer its energy to the one with lower energy.

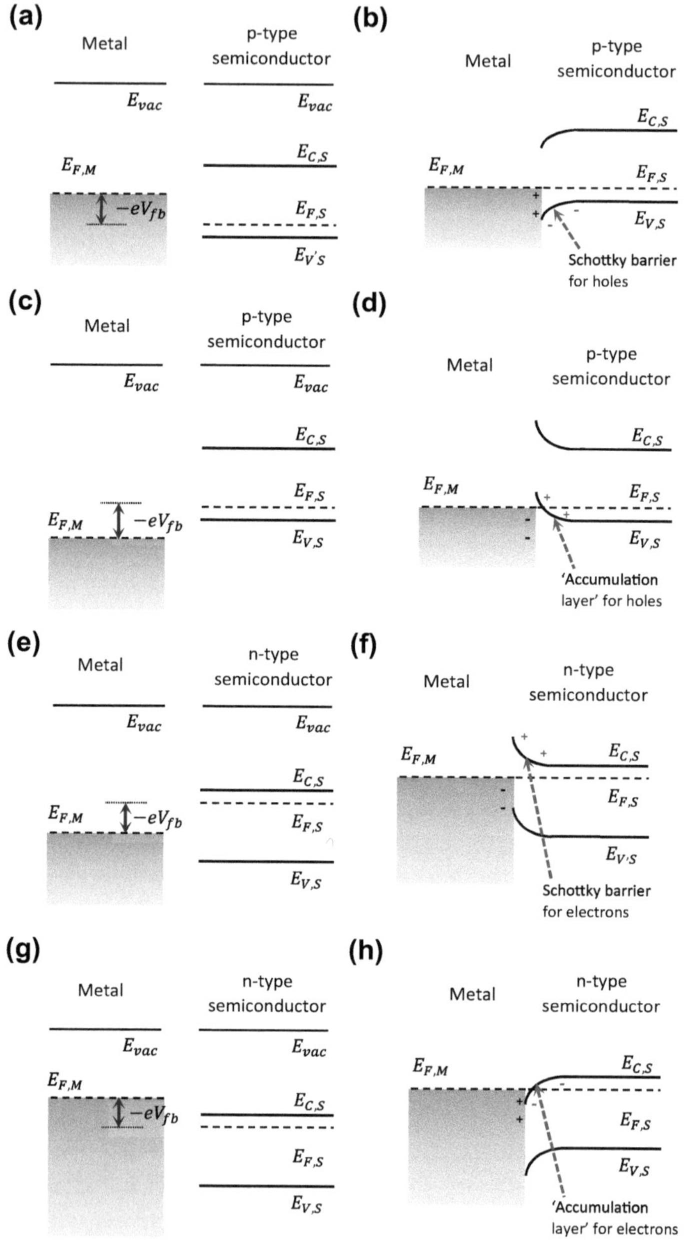

Figure 2.20 Band diagrams of metal–semiconductor heterostructures.

Take an n-type semiconductor as an example. If the Fermi level of a metal sits below the Fermi level of a n-type semiconductor before contact, as shown in Fig. 2.20e, the electrons in the semiconductor will "diffuse" to the metal and stay on its surface. These electrons at the interface would introduce non-zero electric potentials in the metal and the semiconductor. However, we know that the electric fields in metals and in semiconductors is zero. This is because the immobile ionized donors are left in the semiconductor to form a "sheath" (depletion layer). This layer provides an opposite electric potential, and it can screen the electric field from the electrons on the metal.

The screened electric field influences the distribution of charge carriers, and the redistributed carrier density again changes the potential distribution. Therefore, the band is bent at the interface, as depicted in Fig. 2.20f. This depletion layer also serves as the **Schottky barrier** for electrons. This type of metal–semiconductor junction is also called a blocking contact. Note that for a metal–semiconductor junction, an inversion layer cannot be formed. This is because, taking Fig. 2.20f as an example, all holes on the valence band edge at the interface will go uphill until they reach $\mathcal{E}_{F,M}$. Therefore, there are no holes left at the metal–semiconductor interface to form an inversion layer.

Similarly, if the Fermi level of a metal was above the conduction band edge of an n-type semiconductor, an accumulation layer of electrons will form at the interface after contact, as shown in Fig. 2.20g,h. This is sometimes called an **Ohmic contact**.

2.3.3 Field-Effect Transistors

As we are able to engineer the band diagram for heterostructures using different materials, electronic devices with various purposes can be designed, such as diodes, photodiodes, transistors, MOS capacitors, quantum capacitance device, and so on. Transistors are one of the most appealing applications in modern technology as they are essential for both logic circuits, such as CPU, and telecommunication devices, such as high frequency amplifiers. This is mainly due to the ability of the transistor to modulate the electrical current between two of the its contacts by applying an electric field.

The idea of modulating charge carriers using electric field was first introduced to a vacuum triode by L. De Forest in 1906 [Forest (1906)]. This electric field was provided by a metal grid placed between the anode and the cathode of the triode. It acts as a gate for the charge carriers traveling between the anode and the cathode. As the carriers are charged particles, the gating effect can be modulated by the electric field, i.e., the electric field gating effect.

The vacuum triodes were then used in some electrical applications, such as radio amplifiers. However, the development of triodes was limited by the large size and the high operation voltage. In 1930, a patent by J. E. Lilienfeld was issued on the design of a solid-state device used to *modulate carrier density and current*, although no experimental data were presented to support this idea. The aim of this device was to reduce the operation voltage for vacuum triodes. It provides a prototype for the modern transistors, which provide the same functions as the vacuum triodes. In the course of development of transistors, many varieties have been invented, such as bipolar junction transistors, junction field effect transistors, high electron mobility transistors, among others. The gating effect proposed in triodes was then successfully employed to semiconductor transistors.

The metal–oxide–semiconductor field-effect transistor (MOSFET) configuration is one of the most popular designs, as it is low cost and it shows robust performance. In this section we focus on the discussion of MOSFET, readers who are interested in other types of transistors are recommended to consult [Sze (1981); Szkopek (2013)].

It should be noted that the electric field gating effect in transistors is different from that in vacuum tubes. In essence, the field effect in transistors (particularly the metal–oxide–semiconductor field-effect transistors) modulates the charge carriers through the capacitive charging/discharging processes (i.e., engineering the quantum well formed at the oxide–semiconductor interface for charge inversion/depletion). While in triodes, the electric field was adopted as barrier for the electrons traveling between the source an drain. As the intrinsic charge carrier density in semiconductors is low, the carriers induced by the field effect determine the electric current in transistors.

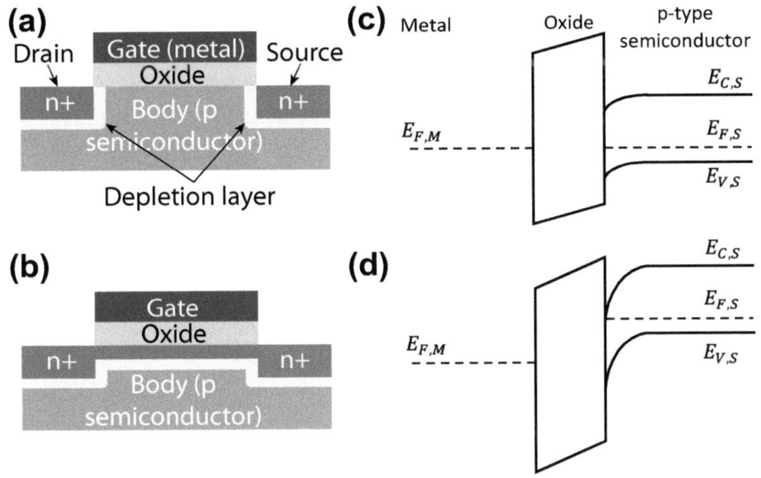

Figure 2.21 Cross-sectional view of a MOSFET at (a) OFF and (b) ON states. (c, d) Band diagrams of the MOS structure corresponding to the states shown in (a) and (b), respectively.

Figure 2.21a illustrates the structure of a typical MOSFET. The electric current flowing between the drain and source can be modulated by the voltage applied on the gate. The drain and source shown in Fig. 2.21a are made of heavily n-doped semiconductor, and they are connected by a moderately p-doped semiconductor body. Note that depletion layers form at the junctions of drain-body and body-source. The gate is usually made of metal which is separated from the body by a thin oxide layer. When there is no voltage applied on the gate, the electric current between the drain and source is 0 due to the existence of the depletion layer. Note that the charge carriers in the body are holes, and the concentration at the oxide–body interface is low, as indicated by the band diagram shown in Fig. 2.21c.

If a positive bias is applied on the gate, an (2D electron gas) is formed at the oxide–body interface, as shown in Fig. 2.21d. The majority charge carrier is then electron, the same as that in the drain and source. This 2D electron gas connects the drain and source, and hence turn the transistor to an ON state, as demonstrated in Fig. 2.21b. Note that this working principle of MOSFET is similar to the idea of ion-sensitive field-effect transistor (ISFET, later discussed in the book), which is a popular potentiometric biosensing technique used to detect chemical species in electrolyte.

In nanoelectronic applications, people are often interested in developing devices with high carrier mobility. This includes the discussion of ballistic limit, scattering, drain-induced barrier lowing, and energy consumption. These topics of MOSFET, however, are not discussed here, as they are more specific for nanoelectronic applications. High carrier mobility in semiconductors is generally not required for applications involving electrolyte, such as biosensors and supercapacitors.

2.4 Summary

This chapter introduced a few important concepts in the classical and modern theories of electron transport, which are summarized as follows.

- In the view of classical theory, the charge carriers in a conductor are scattered by the immobilized atoms/ions. A physical quantity, mobility, is used to describe the scattering effects on the charge transport (Eq. 2.5).
- The Drude model for the DC electrical conductivity (σ) in the absence of magnetic field is $\sigma = qn\mu$, where q, n, and μ are the charge, number density, and mobility of the carriers.
- In the view of quantum mechanics, Fermi–Dirac distribution is used to describe the population of electrons/holes in solid crystals (Eq. 2.9).
- Electrons are waves at the same time in solid crystals. Their motion is described by the Schrödinger's equation.
- Electrons in various nanoelectronic structures are confined in different dimensions (Fig. 2.5).
- The possible states of electrons are quantized, Density of States is then a useful concept to investigate the electrons (Table 2.1).
- Solid crystals have various lattice structures (Fig. 2.9). Reciprocal lattice is a useful tool to investigate electron transport in crystals (e.g., calculate the dispersion relation).

- The crystal structure of lattice introduces bandgaps in the dispersion relation of electrons (Eq. 2.26 and Fig. 2.11).
- The electron properties can be described by an effective mass theory. The electron density is defined using Eqs. 2.29 and 2.30 for common 2D and 3D structures.
- Quantum capacitance is important for electrical devices with very small scale, and it is directly related to the density of states (Eq. 2.35).
- Heterostructures made of various materials are commonly seen in electrical devices.
- Band diagram alignment is a very useful technique to investigate heterostructures (Figs. 2.19 and 2.20).
- Field-effect transistors are essential electrical elements in modern technology including nanoelectronical and biological applications.

Chapter 3

Electrons in Graphene

As discussed in Chapter 1, the electrons in graphene behave differently from those in conventional semiconductors due to the presence of the massless Dirac fermion, the ultrahigh mobility, and the zero bandgap [Castro Neto et al. (2009); Nair et al. (2008)]. These attributes allow the observation of phenomena including the integer quantum Hall effect and the high optical transmittance ($\sim$97%) in the visible range of light. They mainly stem from the unique gapless linear dispersion relation of electrons in graphene and they may be promising for the development of next-generation electronic devices [Semiconductor Industry Association (2009)]. In addition, these unique electronic properties of graphene could be utilized for interpreting and improving the performance of the novel graphene-based electrochemical and photonic applications including biosensors, neural activity recorders, optical diodes, etc.

3.1 Band Structure

The band structure of electrons (or electronic dispersion relation) in a solid crystal determines their transport behavior, as discussed in section 2.2. Graphene is a two-dimensional (2D) crystal which

Graphene–Electrolyte Interfaces: Electronic Properties and Applications
Hualin Zhan

ISBN 978-981-4774-94-9 (Hardcover), 978-1-003-04487-1 (eBook)
www.jennystanford.com

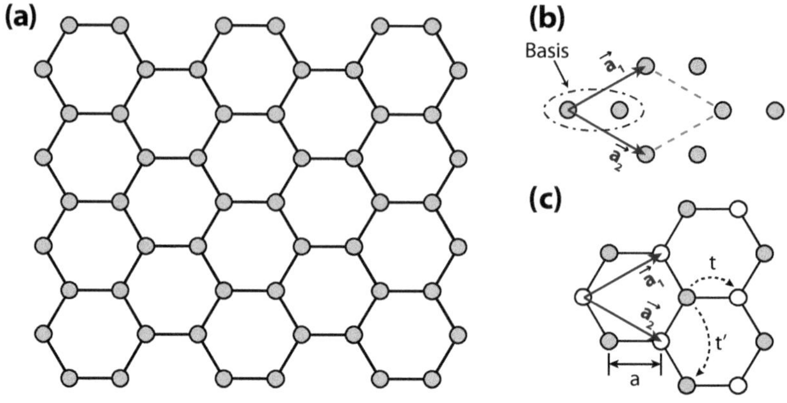

Figure 3.1 (a) The honeycomb atomic structure of graphene. (b) A rhombic primitive cell containing two atoms, where $\mathbf{a}_1$ and $\mathbf{a}_2$ are the primitive translation vectors. (c) The lattice structure consisting of two overlapping triangular sublattices.

possesses unique electrical properties. It is then essential to investigate the crystal structure as well as its influence on the electronic dispersion relation and the density of states.

3.1.1 Crystal Structure and Reciprocal Lattice

Figure 3.1a depicts the **honeycomb** 2D lattice structure of graphene, where the gray dots represent the carbon atoms and the short lines are the bonds. As discussed in section 2.2.1.1, the primitive cell of the lattice can be defined in many different ways, as long as this unit cell can fill the whole lattice space by repetitive translation operations. Figures 3.1b–c show two different representations of the primitive cell. An intuitive thought is to define a rhombic primitive cell (i.e., the area enclosed by two arrows and two dashed lines in Fig. 3.1b) where the basis consists of two atoms (highlighted by the dot-dash ellipse in Fig. 3.1b). $\vec{a}_1 = \frac{a}{2}(3, \sqrt{3})$ and $\vec{a}_2 = \frac{a}{2}(3, -\sqrt{3})$ are the primitive translation vectors. $a \approx 0.124$ nm is the distance between two adjacent carbon atoms. This definition of the primitive cell is straightforward, and it is often used in first-principle numerical calculations of the Schrödinger's equation to obtain the electronic band structure of graphene. It is, however,

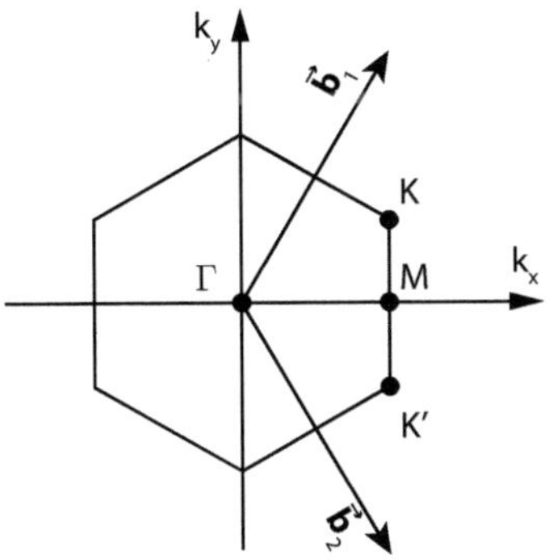

Figure 3.2 First Brillouin zone in the reciprocal lattice of graphene.

more useful to reconstruct the primitive cell for the crystal lattice of graphene such that the analytical expression of the electronic band structure can be obtained. A common method is to introduce two overlapping triangular sublattices (white and gray dots in Fig. 3.1c), so that the electron hopping energy between the adjacent atoms (t) and the next nearest neighbor atoms (t') can be easily included.

The reciprocal lattice is shown in Fig. 3.2, where $\vec{b}_1 = \frac{2\pi}{3a}(1, \sqrt{3})$ and $\vec{b}_2 = \frac{2\pi}{3a}(1, -\sqrt{3})$ are the reciprocal lattice vectors. The positions of K and K′ are $\frac{2\pi}{3a}(1, \frac{1}{\sqrt{3}})$ and $\frac{2\pi}{3a}(1, -\frac{1}{\sqrt{3}})$, respectively.

3.1.2 Dispersion Relation

In the tight-binding Hamiltonian for graphene, electrons are allowed to hop between the nearest neighbor carbon atoms (between white and gray adjacent atoms in Fig. 3.1c) and the next nearest neighbor atoms (among the adjacent atoms in the same sub-lattice, i.e., white or gray), with the energies t and t', respectively. This then allows us to find the dispersion relation in graphene. It has been demonstrated that a general expression of the dispersion relation is written

$$\mathcal{E}_{\pm}(\vec{k}) = \pm t\sqrt{3 + f(\vec{k})} - t' f(\vec{k}), \tag{3.1}$$

where $\vec{k} = k_x\hat{e}_x + k_y\hat{e}_y$ is the wave vector in the reciprocal lattice and

$$f(\vec{k}) = 2\cos(\sqrt{3}k_y a) + 4\cos\left(\frac{\sqrt{3}}{2}k_y a\right)\cos\left(\frac{3}{2}k_x a\right). \tag{3.2}$$

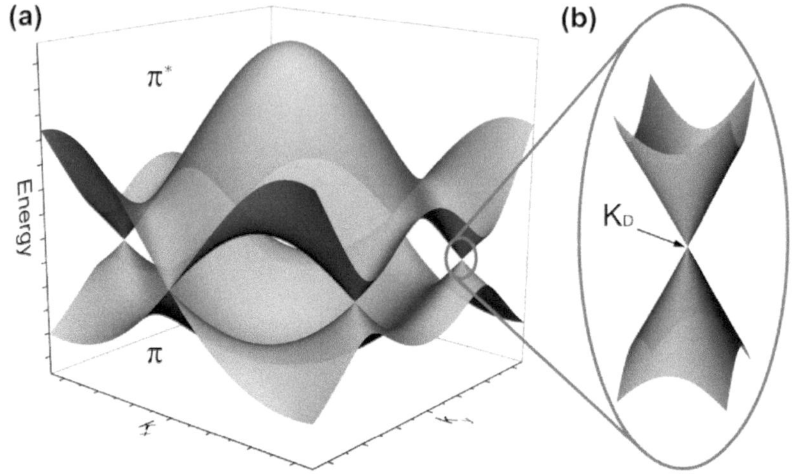

Figure 3.3 (a) 3D plot of the electronic band structure of graphene using a tight-binding approach, where $t = 2.7$ eV and $t' = -0.1t$. (b) Magnification of the band structure close to the Dirac point K_D.

Figure 3.3a shows a 3D plot of the electronic band structure of graphene calculated by Eq. 3.1, where $t = 2.7$ eV and $t' = -0.1t$. The upper and lower parts are denoted as the π^* (conduction) and π (valence) bands. It is interesting to note that the π^* and π bands contact each other at positions such as K and K′. As the Fermi level generally locates at the energy where the π^* and π bands contact, we are then interested in knowing the dispersion relation near this region. Rewriting Eq. 3.1, which includes t' up to the second order approximation, gives

$$\mathcal{E}_\pm(\vec{q}) \approx 3t' \pm \hbar v_F|\vec{q}| - \left[\frac{9t'a^2}{4} \pm \frac{3ta^2}{8}\sin(3\theta_q)\right]|\vec{q}|^2, \qquad (3.3)$$

where $\vec{q} = q_x\hat{e}_x + q_y\hat{e}_y = \vec{k} - \vec{K}$ is the distance of a point in the momentum space to the point K, $\theta_q = \arctan\left(\frac{q_x}{q_y}\right)$, and $v_F = \frac{3}{2}ta \approx 1 \times 10^6$ m/s is the Fermi velocity. This constant Fermi velocity is one of the most striking differences of graphene as compared with usual semiconductors, since the Fermi velocity is dependent on the energy for those semiconductors with parabolic dispersion relation.

It can be seen from Fig. 3.3b that $\mathcal{E}_\pm(\vec{q})$ has a linear relationship to $\vec{q}$ if the second-order approximation term in the right-hand side in Eq. 3.3 is omitted. This generates a cone-like shape at K in a three-dimensional $\mathcal{E}_{\mathrm{k}}$ vs. $\vec{k}$ plot, which is defined as the **Dirac cone**. $\mathrm{K_D}$ is the **Dirac point**. Note that there is **no band gap** shown in Fig. 3.3b. This is a very important phenomenon which distinguishes graphene from the conventional semiconductors. Due to this gapless dispersion relation, graphene is defined as a **semimetal**, as opposed to semiconductor with bandgaps.

It is also demonstrated in Fig. 3.3a that the upper π^* (conduction) and the lower π (valence) bands are asymmetric with respect to zero energy at Dirac point due to the finite value of t'. This asymmetry can be removed if $t' = 0$. Equation 3.3 can then be rewritten as

$$E_{\pm}(\vec{q}) \approx \pm\hbar v_{\mathrm{F}}|\vec{q}| + \mathcal{O}\left[\left(\frac{q}{k}\right)^2\right], \tag{3.4}$$

where $\mathcal{O}\left[\left(\frac{q}{k}\right)^2\right]$ is a second-order approximation term. This relation is the most commonly used **linear** band structure of graphene.

3.1.3 Density of States

Density of states (DOS) quantifies the quantum states for electrons at given energy levels in solid material, which is responsible for describing the electronic behavior. It then can be calculated by counting the number of quantum states per unit volume in momentum space (or the reciprocal lattice), as described in section 2.1.3.3.

Figures 3.4a,c show the DOS of graphene near the position K (or K′) derived from Eq. 3.3, when $t' = 0.2t$ and $t' = 0$, respectively. Similar to the shape of the dispersion relation, there is also no band gap in the DOS. It can be clearly observed in the DOS that the asymmetry between the π^* and π bands can be removed by letting $t' = 0$. In contrast to that shown in Fig. 3.4c, the Dirac point does not locate at the position where $\mathcal{E} = 0$ when $t' \neq 0$ (Fig. 3.4a).

It is then interesting to investigate the DOS near the Dirac point. Figures 3.4b,d show the magnification of the calculated DOS obtained in Figs. 3.4a,c, respectively. As t' is generally a small value depending on the tight-binding parametrization in ab initio calculations [Castro Neto et al. (2009)], we are usually interested in the case where $t = 0$. Using the dispersion relation shown in Eq. 3.4

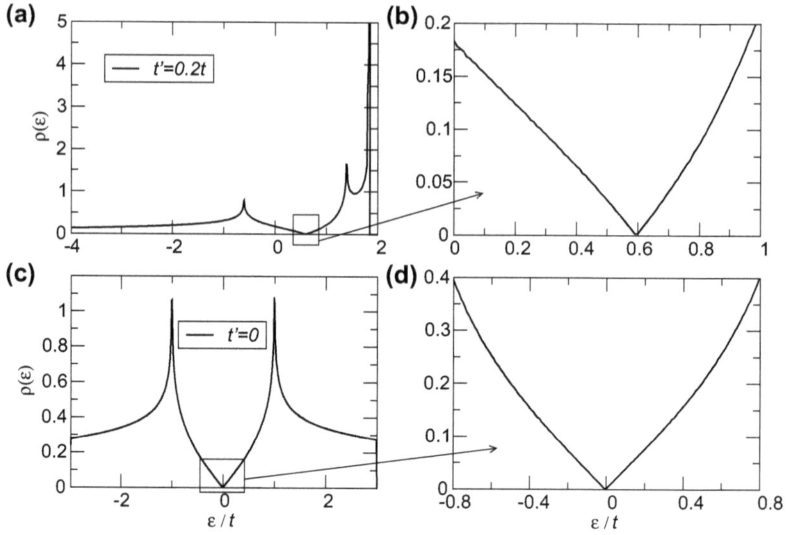

Figure 3.4 Density of states per unit cell as a function of energy when (a, b) $t' = 0.2t$ and (c, d) $t' = 0$. (b) and (d) show the magnification of the density of states close to the Dirac point K_D. Reprinted with permission from [Castro Neto et al. (2009)] Copyright American Physical Society.

and omitting the second-order term, the DOS of graphene per unit area close to the Dirac point can be obtained as

$$\rho(\mathcal{E}) = \frac{\mathrm{d}}{\mathrm{d}\mathcal{E}}\frac{k^2}{\pi} = \frac{2|\mathcal{E}|}{\pi(\hbar v_F)^2}. \tag{3.5}$$

When the electron energy is small (e.g., $|\mathcal{E}| < 0.5t$ in Fig. 3.4d), Eq. 3.5 gives a linear relation between the DOS and the electron energy. This symmetric linear gapless band structure of graphene brings many interesting properties to the charge carriers, such as ultrahigh mobility, very small effective (cyclotron) mass, formation of electron–hole plasma near the Dirac point, symmetric electrical properties for electrons and holes, among others.

3.2 Electronic Properties of Graphene

When studying the electrical behavior of the systems involving graphene, it is useful to know the electrical properties including the charge carrier density and mobility, as they can be used to calculate the electrical conductivity by the Drude model. If the systems involve both graphene and solution, the quantum capacitance of graphene should also be taken into account.

3.2.1 Charge Carrier Density and Doping

As defined in Eq. 2.10, the charge carrier density in graphene is calculated by the integrating the Fermi–Dirac distribution multiplied with the density of states over energy. The discussion in section 2.3.1 indicates that there are two types of charge carriers in graphene (electrons and holes). As shown in Fig. 3.5a, the position of the Fermi level ($\mathcal{E}_\mathrm{F}$) in the DOS-$\mathcal{E}$ diagram of graphene determines the type of the majority charge carriers.

It is then useful to derive the expression of the density of electrons and holes:

$$n_e = \int_0^\infty f_e(\mathcal{E})\rho_e(\mathcal{E})\mathrm{d}\mathcal{E} = N_0 F_1(\eta), \tag{3.6a}$$

$$n_h = \int_{-\infty}^0 f_h(\mathcal{E})\rho_h(\mathcal{E})\mathrm{d}\mathcal{E} = N_0 F_1(-\eta), \tag{3.6b}$$

where $f_e(\mathcal{E}) = \left[\exp\left(\frac{\mathcal{E}-\mathcal{E}_\mathrm{F}}{k_\mathrm{B}T}\right)+1\right]^{-1}$ and $f_h(\mathcal{E}) = 1 - f_e(\mathcal{E}) = \left[\exp\left(\frac{\mathcal{E}_\mathrm{F}-\mathcal{E}}{k_\mathrm{B}T}\right)+1\right]^{-1}$ are the Fermi–Dirac distributions, $\rho_e(\mathcal{E}) = \frac{2\mathcal{E}}{\pi(\hbar v_\mathrm{F})^2} = -\rho_h(\mathcal{E})$ gives the density of states, for electrons and holes, respectively. $\mathcal{E}_\mathrm{F}$ is the Fermi energy or the Fermi level. The Dirac point locates at the position where $\mathcal{E} = 0$. $N_0 = \frac{2(k_\mathrm{B}T)^2}{\pi(\hbar v_\mathrm{F})^2}$. $F_j(\eta) = \frac{1}{\Gamma(j+1)}\int_0^\infty \frac{\xi^j}{\exp(\xi-\eta)+1}\mathrm{d}\xi$ is the **Fermi–Dirac integral** of order j, where $\xi = \frac{\mathcal{E}}{k_\mathrm{B}T}$, $\eta = \frac{\mathcal{E}_\mathrm{F}}{k_\mathrm{B}T}$, and Γ is the gamma function.

Figure 3.5b plots the electron and hole density as a function of the Fermi energy in graphene calculated using Eq. 3.6. The **total carrier density**, $n_t = n_h + n_e$, and the **net charge density**, $n_\mathrm{net} = n_h - n_e$ are shown in Fig. 3.5c. For a graphene/SiO_2/Si device, the net charge density is modulated by the voltage applied on Si using a relation $n_\mathrm{net} = -\alpha V_\mathrm{G}$, where $\alpha \approx 7.2 \times 10^{10}$ cm^{-2}V^{-1} is a geometry-related factor (similar to capacitance) [Novoselov et al. (2005)], and V_G is the gate voltage. It can be seen that when $V_\mathrm{G} > 0$, then

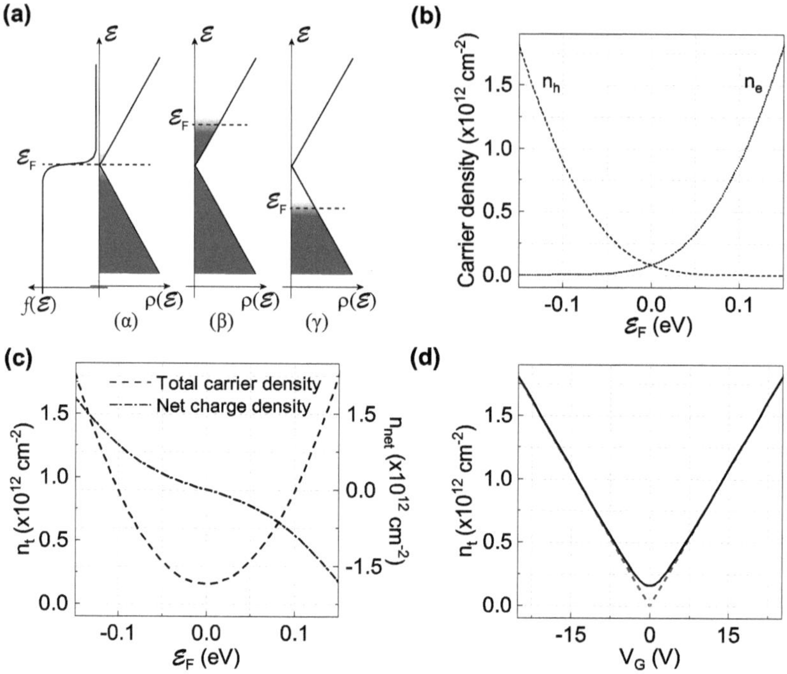

Figure 3.5 (a) The Fermi level ($\mathcal{E}_F$) defined in Fermi–Dirac distribution ($f(\mathcal{E})$) aligned with the density of states of graphene ($\rho(\mathcal{E})$), where the gray and blank regions in $\rho(\mathcal{E})$ represent electrons and holes. α, β, and γ illustrate the cases where the graphene sheet is charge-neutral, electron-rich, and hole-rich, respectively. (b) Electron (dotted line) and hole (short-dashed line) density as a function of the Fermi level. (c) Total carrier density, n_t, (dashed line) and net charge density, n_{net}, (dot-dashed line) as a function of the Fermi level. (d) Total carrier density as a function of the gate voltage of a graphene/SiO_2/Si device (black solid line). The gray dashed lines crossing at 0 are the linear fitting to the charge carrier densities at high gating voltages.

$n_{\text{net}} < 0$, hence electrons accumulate on graphene ($\mathcal{E}_F > 0$), and vice versa. Total carrier density as a function of the gate voltage of such a device is plotted in Fig. 3.5d. Note that the linear fits to the electron and hole densities at high gating voltages cross at 0. This is true when the temperature $T = 0$ K, as Eq. 3.6 can be rewritten as

$$n_e = \int_0^{\mathcal{E}_F} \rho_e(\mathcal{E})\mathrm{d}\mathcal{E} = \begin{cases} 0 & \mathcal{E}_F < 0 \\ \frac{\mathcal{E}_F^2}{\pi(\hbar v_F)^2} & \mathcal{E}_F > 0 \end{cases}, \tag{3.7a}$$

$$n_h = \int_{-\mathcal{E}_F}^{0} \rho_h(\mathcal{E})d\mathcal{E} = \begin{cases} \frac{\mathcal{E}_F^2}{\pi(\hbar v_F)^2} & \mathcal{E}_F < 0 \\ 0 & \mathcal{E}_F > 0 \end{cases}, \tag{3.7b}$$

and hence,

$$n_t = \begin{cases} n_{net} = -\alpha V_G & \mathcal{E}_F < 0 \\ -n_{net} = \alpha V_G & \mathcal{E}_F > 0 \end{cases}. \tag{3.8}$$

However, due to the thermal excitation of electrons at non-zero temperature near the Dirac point, n_t does not reach 0 when $V_G = 0$ due to the coexistence of electrons and holes (Fig. 3.5d).

As shown in Fig. 3.5a, the position of $\mathcal{E}_F$ with respect to the Dirac point in the band structure determines the type and concentration of charge carriers in graphene. $\mathcal{E}_F$ can be shifted by various methods including the direct application of an electrical voltage and doping graphene with different molecules. These dopants can be either chemically bonded with carbon atoms or simply physically adsorbed on the graphene surface. In addition to shifting $\mathcal{E}_F$ due to the charge transfer process between the dopant and graphene, **chemical doping** may introduce asymmetry between the electron and hole mobility, and also change the density of states (which in turn modifies the charge carrier density) [Lherbier et al. (2008); Wehling et al. (2008)].

3.2.2 Quantum Capacitance of Graphene

Quantum capacitance of graphene is calculated by Eq. 2.35, where the density of states is represented by Eq. 3.5,

$$C_q = e^2\rho(\mathcal{E}_F) = \frac{2e^2|\mathcal{E}_F|}{\pi(\hbar v_F)^2}, \tag{3.9}$$

as indicated by the solid black line in Fig. 3.6a. Note that C_q changes linearly with $\mathcal{E}_F$. The empty gray squares in Fig. 3.6a describe the capacitance directly derived from the total carrier density, which are consistent with the result of Eq. 3.9. However, since the net charge carrier density is regularly measured in experiments, as opposed to the total carrier density, the differential quantum capacitance (C_q) values derived from n_{net} are represented by the gray empty circles in Fig. 3.6a. Therefore, the non-zero smooth transition of C_q is observed in experiments [Ponomarenko et al. (2010)]. It should be noted that

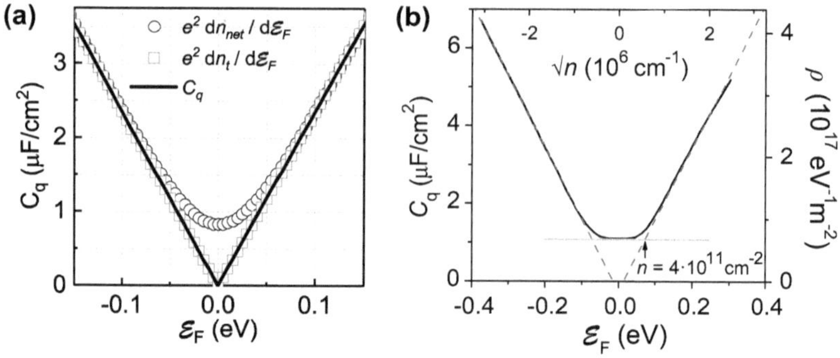

Figure 3.6 (a) Calculated quantum capacitance by Eq. 3.9 using DOS (solid black line), $e^2 \left| \frac{dn_{\text{net}}}{d\mathcal{E}_F} \right|$ (gray empty circles), and $e^2 \left| \frac{dn_t}{d\mathcal{E}_F} \right|$ (gray empty squares). (b) Measured quantum capacitance (black line) as a function of the Fermi level $\mathcal{E}_F$. Reprinted with permission from [Ponomarenko et al. (2010)]. Copyright American Physical Society.

the experimental methods and the impurities also contribute to this phenomenon.

Figure 3.6b shows the quantum capacitance measurement of a graphene capacitor, where a very thin aluminum oxide layer (1 nm), formed by depositing aluminum in the presence of oxygen, was used as the dielectric layer between a graphene sheet and an aluminum electrode. The black line in Fig. 3.6b is the measured quantum capacitance, where the gray dashed lines are the extrapolation (fitted) lines. The plateau in the measured C_q curve results from the nonlinear stretching of the x axis, as $\mathcal{E}_F$ is calculated from the measured gate voltage. As can be seen from Fig. 3.6b, the quantum capacitance of graphene is indeed linear with respect to the Fermi level, as predicted by the theory. This conclusion is useful for electrical measurements of graphene in liquid, as the double layer capacitance (explained in Chapter 4) generally has a large value.

3.2.3 Mobility and Scattering

In the study of charge carrier transport, the Drude model of electrical conductivity (section 2.1.1.2) suggests that the mobility (μ) of the carriers is also very important. In semiconductors, $\mu = \frac{q\tau}{m^*}$ is defined in the effective mass theory (section 2.2.1.2), where $m^* = m\left(1 \pm \frac{4E_k}{E_g}\right)^{-1}$ is the effective mass. As the effective mass of electrons (where m is the classical electron mass, E_k is the kinetic energy of an unscattered electron (wave), and E_g is the bandgap, as described in Eq. 2.26) is related to the bandgap (the smaller the bandgap, the lighter the effective mass), μ is then expected to have a very high value for graphene.

3.2.3.1 Mobility

In contrast to a conventional semiconductor which has a band gap and a parabolic relation in the band structure, the dispersion relation of graphene is gapless and linear. This feature gives an ultra-high mobility and saturation velocity of electrons in graphene. The gapless feature ($E_g = 0$) produces a massless fermion (electron), and this in turn gives a very high mobility. The comparison of the electron mobility in graphene with other semiconductors is plotted in Fig. 3.7a. It is shown in the figure that the mobility in graphene is an order of magnitude higher than the other materials. For graphene nanoribbons (GNRs), the mobility of the charge carriers is a function of the ribbon width, as indicated by the simulation and experimental results in Fig. 3.7b.

In the presence of an electric field, electrons in semiconductor are accelerated. The velocity eventually reaches a saturation limit

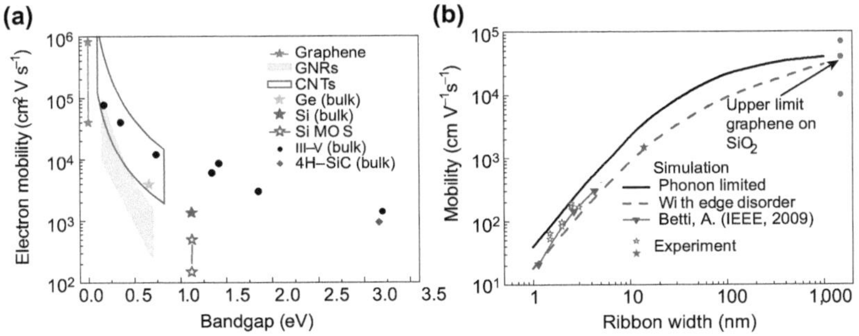

Figure 3.7 (a) The electron mobility vs. the bandgap for different materials. (b) Charge carrier mobility of graphene nanoribbons as a function of the ribbon width. Reprinted with permission from [Schwierz (2010)]. Copyright Nature Publishing Group.

in nanoscale devices [Szkopek (2013)]. This is strongly material dependent. In theory, electrons in graphene have a very high saturation velocity. This is limited by the optical phonons and ultimately restricted by the slope of the linear dispersion relation (i.e., the Fermi velocity), although there has no convincing experimental evidence been reported yet [Schwierz (2010)].

3.2.3.2 Scattering

It can be seen from the definition of mobility that the relaxation time (τ) induced by scattering also plays an important role. In general, carriers in semiconductors are scattered by phonons[1] and charged impurities. Additionally, in graphene, the ripples introduced by the interaction between graphene and the substrate could also have an effect. However, studies show that the influence of phonons and ripples on the mobility or conductivity in graphene is very small at room temperature [Chen et al. (2008b,c); Hwang et al. (2007a)]. Therefore, scatterings by phonons and the ripples in graphene are not discussed in this section. It is then important to understand how the electrons in graphene interact with the charged impurities. Previous theoretical and experimental works have demonstrated that *the scattering of electrons/holes in graphene is dominated by the short-range collision and long-range Coulomb interaction (**charged impurity scattering**) at high and low carrier density*, respectively [Adam et al. (2007); Chen et al. (2008b); Hwang et al. (2007a); Martin et al. (2008)]. This is because the mean free path[2] is determined by $\lambda_s \sim 1/\sqrt{n}$ and $\lambda_s \sim \sqrt{n}$ for short-range and long-range scatters, respectively [Adam et al. (2007)].

In low carrier concentration, the impurities in the environment introduce electron–hole puddles in graphene, as the 2D nature makes graphene highly sensitive to the environment. These puddles in turn scatter electrons/holes, and therefore changes the carrier mobility. The electron–hole puddles have been observed by a

---

[1] Phonons are the quanta of lattice oscillations in crystal, which is similar to the idea that photons are the quanta of electromagnetic (including light) waves [Davies (1998); Szkopek (2013)].

[2] The average distance between two collision events when the electrons travel in the crystal [Kittel (2004)].

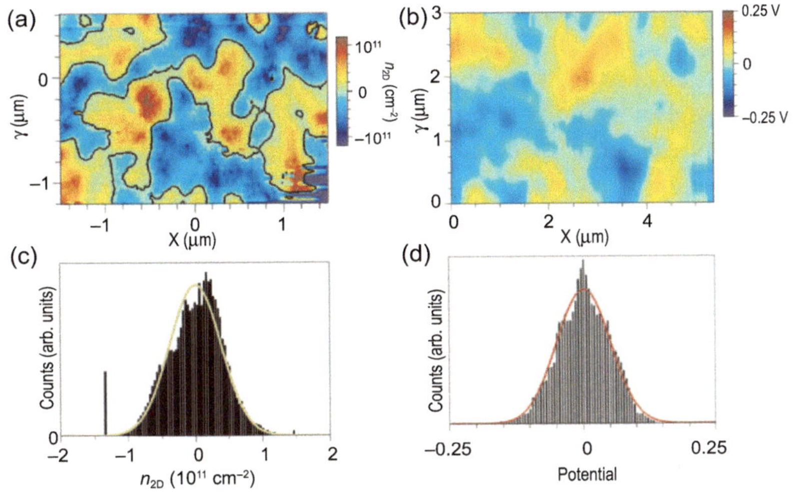

Figure 3.8 Color maps of (a) electron–hole puddles (blue: holes, red: electrons) and (b) the spatial fluctuation of the measured surface potential of a graphene sample, respectively. (c) Histogram of the density distribution in (a). (d) Histogram of the fluctuation which follows Gaussian distribution. Reprinted with permission from [Martin et al. (2008)]. Copyright Nature Publishing Group.

scanning single-electron graphene transistor [Martin et al. (2008)]. Figure 3.8a shows the observed electron–hole puddles in graphene, while the measured spatial fluctuation of the surface potential indicated in Fig. 3.8b experimentally suggest that the fluctuation induced by the charged impurities is the cause of the formation of the electron–hole puddles.

In addition, the paper suggests that both the electron–hole puddle density and the surface potential fluctuation follow a Gaussian distribution, as shown in Figs. 3.8c,d. Therefore, an empirical expressions for carrier density can be written as

$$\begin{aligned} n^*(\mathcal{E}_\mathrm{F}) &= \sum_{\mathcal{E}_\mathrm{F}} n(\mathcal{E}_\mathrm{F}) P(\mathcal{E}_\mathrm{F}) \\ &= \frac{1}{\sqrt{2\pi}\sigma} \int_{-\theta}^{\theta} n(\mathcal{E}_\mathrm{F}^*) \cdot \exp\left[-\frac{(\mathcal{E}_\mathrm{F}^* - \mathcal{E}_\mathrm{F})^2}{2\sigma^2}\right] \mathrm{d}\mathcal{E}_\mathrm{F}^*, \end{aligned} \tag{3.10}$$

where $\mathcal{E}_\mathrm{F}^*$ represents the local Fermi level perturbed by the impurities in the environment. θ represents the integration limits, σ

is the standard deviation, $n(\mathcal{E}_F^*)$ is the local carrier density described by Eq. 3.6, and $P(\mathcal{E}_F)$ is the probability function which follows Gaussian statistics. Note that all the electron-hole puddles within the integration limits (from $-\theta$ to θ) are considered as charged carriers, the empirical equation 3.10 could then be used to calculate the charge carrier density in the presence of charged impurities.

3.3 Nanoelectronic Applications

Owing to the advantage of high carrier mobility, graphene is concluded suitable for the development of the next-generation electronic devices [Semiconductor Industry Association (2009)]. Transistors are one of the most appealing applications as they are essential for both logic circuits, such as CPU, and telecommunication devices, such as high-frequency amplifiers.

3.3.1 Graphene Field-Effect Transistors

The metal–oxide–semiconductor field-effect transistor (MOSFET, discussed in section 2.3.3) is one of the most popular configurations in modern technology considering its low cost and the robust performance. As the charge carriers in graphene can be modulated by varying the externally applied electric field (i.e., field effect), it is then possible to make field-effect transistors using graphene. Due to its unique 2D atomic structure, a graphene field-effect transistor can have either a back-gate or top-gate configuration, as shown in Figs. 3.9a and b, respectively. In both cases, the gate electrode is separated from graphene by an oxide layer. Both of these configurations are commonly used in nanoelectronic researches on graphene.

As the charge carriers transport rapidly in graphene due to the high mobility, the logic transistors respond quicker and the cut-off frequency of the high-frequency (HF) transistors increases [Schwierz (2010); Szkopek (2013)]. However, although the gapless feature of the band structure in graphene gives excellent mobility, it causes a possibly fatal problem for the logic transistors. Since

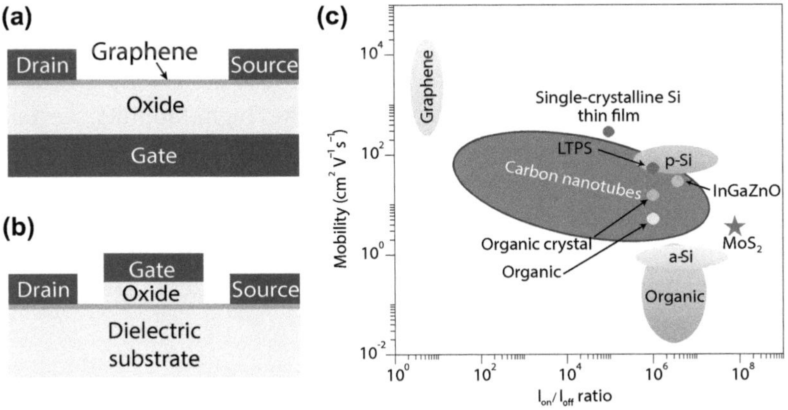

Figure 3.9 (a) The back-gate and (b) top-gate configurations for graphene field-effect transistors. (c) A map displaying the trade-off between the electron mobility and $I_{\text{on}}/I_{\text{off}}$ ratio for field effect transistors made of different materials. Reprinted with permission from [Fiori et al. (2014)]. Copyright Nature Publishing Group.

the transistors made of graphene cannot be switched off due to the absence of bandgap, the leakage current (I_{off}) for a CPU consisting of billions of transistors will be significant. Therefore the power dissipation will be huge with a very small on–off current ratio $I_{\text{on}}/I_{\text{off}}$. For the next-generation logic transistors, $I_{\text{on}}/I_{\text{off}}$ needs to be larger than 10^4, while for graphene logic transistors have a value of approximately 100, as shown in Fig. 3.9c. A resolution to this problem is to open a band gap for graphene, such as fabricate a graphene nanoribbon or a bilayer graphene [Schwierz (2010)]. However, the complicated electronic structure induced by graphene edges complicates this idea. HF transistors on the other hand do not require to be switched off, since their main function is to amplify the signal. The cut-off frequency of a graphene MOSFET ($f_{\text{T}} = \frac{g_m}{2\pi C} = \frac{v_{\text{F}}}{\pi^2 L}$) which depends on the effective transit rate of electrons v_{F} across the channel length L is almost comparable to high electron mobility transistor (HEMT) due to its high mobility and saturation velocity. However, since the cut-off frequency should be operated in the saturation regime, and graphene MOSFETs have not reached the saturation point yet, the device speed is disadvantaged.

These disadvantages suggest that more research needs to be carried out to develop a commercialized graphene nanoelectronic device. These limitations, however, do not affect the applications of graphene involving electrolyte, as the high mobility, certain value of I_{on}/I_{off}, and/or high saturation velocity are generally not required. The gapless feature of the dispersion relation and the 2D atomic lattice structure make graphene very sensitive to its environment. This is of interest to the applications including biology and energy storage, which will be discussed in detail later in this book.

3.3.2 Quantum Capacitance Devices

In nanoelectronics, graphene is regarded as a 2D material, not only due to its atomic crystal structure, but also because it serves as a quantum well which only allows electron gas move in 2D. As discussed in sections 2.2.3.2 and 3.2.2, the Pauli exclusion principle results in quantum capacitance for two-dimensional electron gas (2DEG). This gives rise to a partial penetration effect of the electric field when a 2DEG is incorporated into a system where the quantum capacitance must be taken into account [Luryi (1988)]. This is because the total energy of the system must stay in minimum at equilibrium.

Vertical graphene tunneling field-effect transistor (TFET) is one of the applications which takes the advantage of the partial penetration effect due to quantum capacitance [Georgiou et al. (2013)]. Figure 3.10a demonstrates the design of such a vertical TFET, where two graphene sheets (Gr_B and Gr_T) are separated by a few-layered tungsten disulfide (WS_2) barrier. All these thin films are placed on the top of a $Si/SiO_2/BN$ structure where Si is utilized as the gate. A cross-section high-resolution high-angle annular dark-field scanning transmission electron microscopy (HAADF STEM) image of a fabricated device is shown in Fig. 3.10b.

In contrast to the traditional TFETs made of heavily doped semiconductors where the charge carriers tunnel through the p–n junctions [Ionescu and Riel (2011)], the charge carriers in graphene sheets tunnel through the thin WS_2 film barrier in the vertical

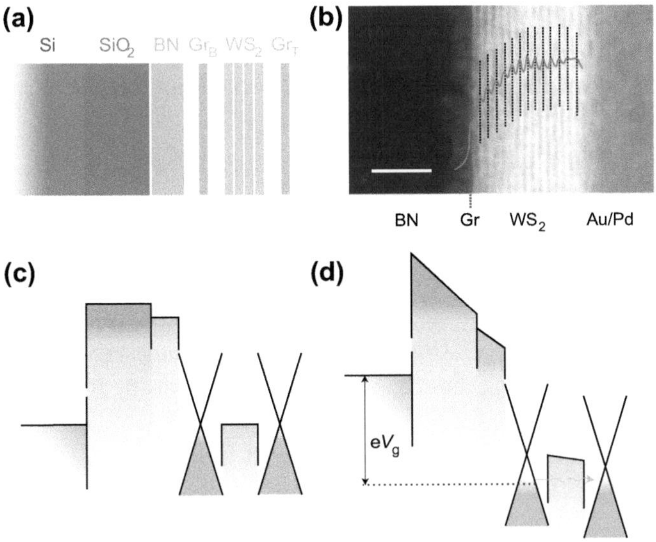

Figure 3.10 (a) A schematics for the design of a flexible tunneling field-effect transistor (TFET) made of graphene. (b) A cross-section high-resolution high-angle annular dark-field scanning transmission electron microscopy (HAADF STEM) image of a fabricated TFET (scale bar: 5 nm). (c, d) Band diagrams of the device when $V_g = 0$ (c) and $V_g \neq 0$ (d), respectively. Reprinted with permission from [Georgiou et al. (2013)]. Copyright Nature Publishing Group.

TFET. As shown in Fig. 3.10c, when there is no gate voltage applied ($V_g = 0$), the charge carriers' concentration in both Gr_B and Gr_T stays at minimum, as the Fermi level is positioned as the Dirac point. When a non-zero gate voltage is applied ($V_g \neq 0$), the charge carriers concentration in both Gr_B and Gr_T increase due to the field-effect and the partial penetration effect of electric field on Gr_B (Fig. 3.10d). This would drastically enhance the tunneling current, and then provide a different route to energy-efficient nanoelectronic devices. Although this device is made of "dry" graphene sheets without the inclusion of solutions, the concepts of quantum capacitance and the partial penetration effect of electric field as well as their applications are useful to graphene–electrolyte systems.

3.4 Summary

This chapter applies the knowledge from Chapter 2 to graphene. Since graphene is a unique 2D gapless semiconductor (semimetal), some of its peculiar properties are summarized as follows.

- Graphene has a honeycomb 2D lattice structure where the interatomic distance is 0.124 nm (Fig. 3.1).
- K and K′ in the reciprocal lattice are defined as the Dirac point, where the π^* and π bands contact, i.e., no bandgap.
- The dispersion relation near the Dirac points is linear if the electron hopping energy between the next nearest neighbor atoms is 0 (Eq. 3.4).
- Density of states is linear with respect to the energy ($\mathcal{E}$), and it is symmetric about $\mathcal{E} = 0$ (Eq. 3.5 and Fig. 3.4).
- The zero bandgap and thermal excitation indicate that graphene contains both electrons and holes even when the Fermi level is positioned at the Dirac point at room temperature.
- Quantum capacitance of graphene is generally proportional to the density of states. However, the exact shape of the quantum capacitance depends on the measurement technique (Fig. 3.6).
- The carrier mobility of graphene is among the highest ones at present. However, for graphene nanoribbons, the mobility decreases with the width of the ribbon.
- The scattering of electrons/holes in graphene is dominated by the short-range collision and long-range Coulomb interaction (charged impurity scattering) at high and low carrier density, respectively.
- The electron–hole puddles in graphene, which induced by the charged impurities, follow a Gaussian distribution, and they contribute to the residual conductivity at zero gate voltage.
- Graphene can be used to make field-effect transistors. However, the low $I_{\text{on}}/I_{\text{off}}$ limits its future development for logic electronic applications.
- Partial penetration effect of electric field and the quantum capacitance of graphene are useful for energy-efficient electronic devices and graphene–electrolyte systems.

Chapter 4

Electrons in Electrolyte

In contrast to the semiconductor (including graphene) which is an electron/hole conductor, electrolyte is an ionic conductor. The electrons in semiconductor follow the Fermi–Dirac distribution [Kittel (2004)], while the ions in solution follow the Maxwell–Boltzmann statistics [Pathria and Beale (2011)]. The physical processes on the semiconductor–electrolyte interface, including charge transfer and the establishment of local electric field, are then relevant to both systems. Since the electrons are exchanged between the solid state lattice and the mobile ions in solution at the interface during a charge transfer process, it is important to investigate the electronic properties in both graphene and the ions in solution. In addition, electrons could be induced in semiconductors by the presence of ions on the semiconductor-electrolyte interface, where the density, mobility, and band structure of electrons would in turn affect ions. Electrochemistry is a branch of physical chemistry which studies the electrical processes at electrode–electrolyte interfaces. Therefore, we must review the fundamentals of electrochemistry before the investigation of graphene–electrolyte systems.

Graphene–Electrolyte Interfaces: Electronic Properties and Applications
Hualin Zhan

ISBN 978-981-4774-94-9 (Hardcover), 978-1-003-04487-1 (eBook)
www.jennystanford.com

4.1 Elementary Theories

In order to understand the kinetics of electrons at semiconductor–electrolyte interfaces, it is useful to first study the ion dynamics in solution. Hence, a review of the fluid mechanics is presented in the beginning of this section for the investigation of the ion dynamics. Since we are interested in the electronic behavior at graphene–electrolyte interfaces in this book, it is then important to revisit the well-known theories for electrons at solid–electrolyte interfaces established by R. A. Marcus [Marcus (1956)] and H. Gerischer [Gerischer (1990)].

4.1.1 The Fluid Mechanics

The classical fluid mechanics uses a set of equations obtained by taking the moments of the Boltzmann transport equation (appendix A.3) to describe the dynamical behavior of an infinitesimal group of particles (**fluid parcel**) in position space (either Lagrangian or Euler coordinates). These equations can be adopted for the discussion of a range of different fluids including gas, liquid (electrolyte), plasma, and even electron gas in solid materials in some cases.

4.1.1.1 The Nernst–Planck equation

It is usually difficult to obtain a general solution for the fluid equations. Therefore, a common method to study the dynamics in fluid is to impose some assumptions. In most cases, we are interested in the electrolyte at steady states where the momentum of the fluid parcel does not change over time. Hence the left-hand side (LHS) of Eq. A.13 vanishes, and the equation becomes

$$0 = -\nabla p + n\vec{F}, \tag{4.1}$$

where $p = nk_{\mathrm{B}}T$ is the pressure, n is the ionic concentration, k_{B} is the Boltzmann constant, T is the temperature, and $\vec{F}$ represents the volumetric force applied on the fluid parcel. The viscosity and the strain rate tensor are neglected for simplicity.

Note that the derivation process of Eq. 4.1 has neglected the contribution from collision between different ions in the electrolyte.

However, the collision processes are not insignificant in most cases of interest in electrolyte. It is then important to include the collision term in Eq. 4.1. As discussed in section 2.1.1.1, the BGK operator is a commonly used model for the collision term, thus Eq. 4.1 can be rewritten as

$$0 = -\nabla p + n\vec{F} - \frac{nm\vec{u}}{\tau}, \tag{4.2}$$

where m is the ion mass, $\vec{u}$ is the mean velocity for the fluid parcel, and τ represents the relaxation time (i.e., the average time duration between collision events).

When there is an electric field present in the electrolyte, the volumetric force on an ion is $\vec{F} = q\vec{E}$, where q represents the charge of the ion and $\vec{E} = -\nabla\varphi$ is the electric field strength. φ is the electric potential in the electrolyte. If we consider that the spatial distribution of the temperature is uniform in the system (i.e., $\nabla T = 0$), Eq. 4.2 can be rewritten as

$$\vec{\Gamma} = n\vec{u} = -D\nabla n - n\mu\nabla\varphi, \tag{4.3}$$

where $\vec{\Gamma}$ represents the flux, $D = k_{\mathrm{B}}T\frac{\tau}{m}$ is the diffusion coefficient, and $\mu = q\frac{\tau}{m}$ is the mobility, of the ion in the electrolyte. The **Einstein relation** can be obtained:

$$\frac{D}{\mu} = \frac{k_{\mathrm{B}}T}{q}. \tag{4.4}$$

In addition to the application of Eq. 4.3 in liquid, this expression is also used in plasma physics by the name of **ambipolar diffusion** as the charges with opposite signs transport differently. This equation describes that the net flux of a charged species in fluid should take into account both the diffusion driven by the concentration gradient and the migration induced by electric field.

On some occasions where the bulk electrolyte drifts in a certain direction with the velocity $\vec{v}$, Eq. 4.3 becomes

$$\vec{\Gamma} = -D\nabla n - n\mu\nabla\varphi + n\vec{v}. \tag{4.5}$$

This is the famous **Nernst–Planck equation**, which is commonly used in describing the ion dynamics in solution. This equation, or its simplified form (Eq. 4.3), is generally combined with Eq. A.11 and Poisson's equation to investigate the temporal and spatial evolution of ions, namely, the **Poisson–Nernst–Planck** model.

4.1.1.2 Electrochemical potential

In electrolyte, there are typically anions and cations with negative and positive charges, respectively. This is similar to the case in plasma where electrons and positive ions have the opposite charges. Therefore, the discussion for ambipolar diffusion from plasma physics can also be adopted to describe the ion dynamics in electrolyte, as Eq. 4.3 is suitable for both.

If we consider the drift velocity $\vec{v} = 0$ in Eq. 4.5, the net electrical current in an electrolyte can be written as

$$\begin{aligned} \vec{J} &= q_a\vec{\Gamma}_a + q_c\vec{\Gamma}_c \\ &= q_a(-D_a\nabla n_a + n_a\mu_a\vec{E}) + q_c(-D_c\nabla n_c + n_c\mu_c\vec{E}), \end{aligned} \tag{4.6}$$

where the subscripts a and c denote the anions and cations in the electrolyte (or the electrons and positive ions in plasmas), respectively.

When there is no net electrical current in the electrolyte, i.e., $\vec{J} = 0$, Eq. 4.6 can be rearranged,

$$\vec{E} = \frac{k_\mathrm{B}T\,(\mu_c\nabla n_c + \mu_a\nabla n_a)}{q_c n_c\mu_c + q_a n_a\mu_a} = -\nabla\varphi, \tag{4.7}$$

where $\vec{E}$ represents the ambipolar field. Note that the Einstein relation (Eq. 4.4) has been used. Eq. 4.7 can be rewritten as $\mathrm{d}\Psi(n_c, n_a) = \theta\mathrm{d}\varphi = -\theta\left[\frac{k_\mathrm{B}T\,(\mu_c\mathrm{d}n_c+\mu_a\mathrm{d}n_a)}{q_c n_c\mu_c+q_a n_a\mu_a}\right]$, where θ is an integrating factor. This is an exact differential equation if $\theta = \frac{q_c n_c\mu_c+q_a n_a\mu_a}{n_c\mu_c+n_a\mu_a}$. Thus,

$$\varphi - \varphi_0 = -k_\mathrm{B}T\int\frac{\mathrm{d}(n_c\mu_c + n_a\mu_a)}{q_c n_c\mu_c + q_a n_a\mu_a}. \tag{4.8}$$

For semiconductors with only one type of non-degenerate carrier gas (i.e., electron or hole gas) and classical plasmas, the masses of electrons and ions greatly differ, $m_a \ll m_c$, hence $\mu_a \gg \mu_c$. Note that this assumption is valid for some electrolytes as well. Integration of Eq. 4.8 yields

$$\varphi - \varphi_0 = -\frac{k_\mathrm{B}T}{q_a}\ln\left(\frac{n_a}{n_{a0}}\right), \tag{4.9}$$

and hence,

$$n_a = n_{a0}\exp\left(-\frac{q_a(\varphi-\varphi_0)}{k_\mathrm{B}T}\right) = n_{a0}\exp\left(-\frac{\mathcal{E}-\mathcal{E}_\mathrm{F}}{k_\mathrm{B}T}\right), \tag{4.10}$$

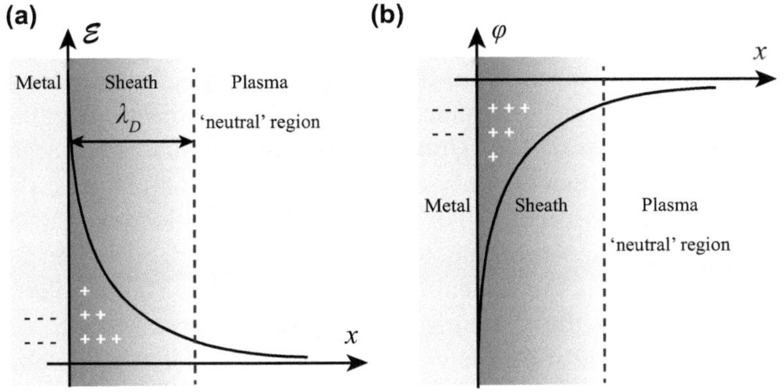

Figure 4.1 (a) Electron energy profile, and (b) potential distribution at a metal–plasma interface.

where

$$\mathcal{E} = \bar{\mu}_\text{p} = \mu_\text{p} + q_a\varphi, \tag{4.11}$$

$$\mathcal{E}_\text{F} = \mu_\text{p} + q_a\varphi_0, \tag{4.12}$$

μ_p is the chemical potential. $\bar{\mu}_\text{p}$ is the **electrochemical potential**, which has a similar form to the Fermi energy (Eq. 4.12). Note they have the meaning of free energy.

Equation 4.9 describes the potential distribution at the solid–fluid interface, which is also the **Nernst equation** in electrochemistry. Figure 4.1 shows the energy and electrical potential profiles at a metal–plasma interface as an example. Note that the electron energy curve shown in Fig. 4.1a always have an opposite sign as compared with the potential (shown in Fig. 4.1b) since $\mathcal{E} \propto q_a\varphi = -e\varphi$. Equation 4.10 describes the non-degenerate 3D electron gas distribution in semiconductors, namely the Maxwell–Boltzmann distribution.

4.1.1.3 Debye screening

In ambipolar diffusion, charged particles are re-distributed within a shallow layer in plasma at the metal–plasma interface. This layer is the so-called plasma sheath where λ_D is the Debye length which can be used to represent the thickness of the sheath to an extent, as shown in Fig. 4.1. This concept can be applied to liquid electrolytes where the charged particles are cations and anions, except that an electrical double layer, rather than plasma sheath, is formed. **Debye length** is defined as the distance at which the potential perturbation caused by an external field/object is screened. When the charged particles follow Boltzmann's statistics in liquids or plasmas, this effect is Debye screening. For the particles (i.e., electrons/holes) in heavily doped semiconductors which are described by Fermi–Dirac distribution, the screening effect is defined as Thomas–Fermi screening, where the screening length is Thomas–Fermi length.

The magnitude of sheath width is usually several times the Debye length λ_{D}. Substituting the Boltzmann distribution, i.e., Eq. 4.10, into Poisson's equation, one obtains

$$\lambda_{\mathrm{D}} = \sqrt{\frac{\epsilon_0 k_{\mathrm{B}} T}{n_{a0} q_a^2}}, \tag{4.13}$$

where ϵ_0 is the permittivity of vacuum if the fluid is plasma. For electrolyte, ϵ_0 is the dielectric constant of the solution. According to the definition of λ_{D}, the charged particle density at λ_{D} is $n_a = n_{a0}\exp(-1)$. Substitution of Eq. 4.9 into Eq. 4.13 gives $\lambda_{\mathrm{D}} = W_{\mathrm{dep}} = \sqrt{\frac{\epsilon_0(\varphi-\varphi_0)}{n_{a0} q_a}}$. Note that this is the same as the commonly accepted depletion layer thickness (W_{dep}) of semiconductor except there is a $\sqrt{2}$ difference. This difference comes from the approximation used in semiconductor physics where no electrons (holes) are left in the depletion layer, i.e. the donors (acceptors) are distributed uniformly within the layer.

From these derivations, we confirm that there are similarities in charge transport phenomena among moderately doped semiconductor, electrolyte and plasmas. These similarities originate from the Boltzmann distribution of the particles. This analogy facilitates the study of electronic behavior at semiconductor–electrolyte interfaces, especially energy band alignment.

Charges in metal and heavily doped semiconductors, on the other hand, are degenerate and are described by Fermi–Dirac distribution. Thus the degenerate electron gas should not be described by Eq. 4.10. It is shown that the Thomas–Fermi screening length is written as $\lambda_{\mathrm{TF}} = \sqrt{\frac{2}{3}\frac{\epsilon \mathcal{E}_{\mathrm{F}}}{n_{a0} q_a^2}}$ [Szkopek (2013)]. However, since the charge density in metal is very high, an estimation of the screening length gives a very small number compared to λ_D of electrolyte or W_{dep} of moderately doped semiconductor. This indicates that the charged carriers are accumulated on the surface of the metal. This conclusion is very useful in energy band alignment (using the Gerischer model) and capacitance estimation (for non-Faradaic processes).

4.1.2 Marcus Theory for Electron Transfer

In electrochemistry, we are generally interested in knowing how the electrons transfer between the electrode and electrolyte. As discussed previously, the fluid mechanics is useful for the study of ion dynamics in electrolytes (both in the bulk and at the electrode–electrolyte interface). It describes an ensemble of ionic species in liquid, where the energy of the entire species is included in the distribution function. However, we are often more interested in investigating the nature and the structure of the reacting species in the molecular level to understand the kinetics of the electron transfer processes. R. A. Marcus proposed a microscopic model in 1956 to explain the rates of electron transfer reactions, and he received the Nobel Prize in chemistry in 1992 for this contribution.

When a chemical species receives/loses electrons from the environment (such as an electrode or another chemical species in the solution), it is reduced/oxidized. This process is regarded as a reduction/oxidation (**redox** for short) reaction described as

$$O + e^- \rightleftharpoons R, \tag{4.14}$$

where O and R are the oxidized and reduced chemical species, and e^- represents the electrons being transferred. The right/left arrow denotes the reduction/oxidation process.

It is worth mentioning that the solvent molecules in an electrolyte (which are from the background liquid, such as the water molecules in aqueous solutions) intend to stay in the close proximity of the chemical compounds (e.g., ions), and form an interfacial sphere, i.e., **solvation shell** (or hydration shell for that in aqueous solutions). An ion with the solvation shell is called a solvated ion.

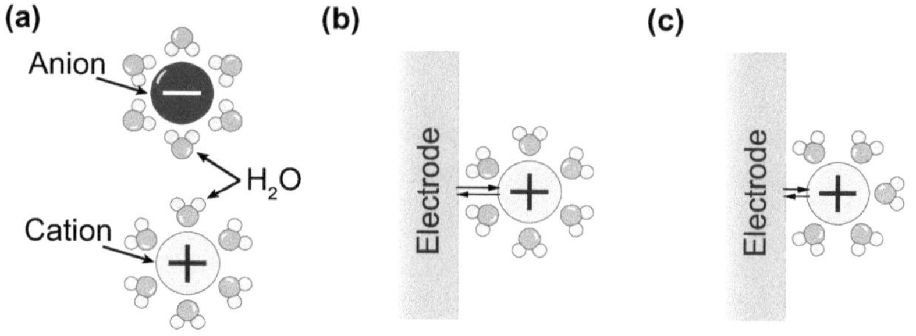

Figure 4.2 (a) Solvated anion and cation with a hydration shell. (b) Outer-sphere and (c) inner-sphere electron transfer processes between a solvated cation and the electrode.

Figure 4.2a shows a solvated anion and cation with a hydration shell. Note that the energy involved in the reorganization of the solvation shell is defined as the reorganization energy (λ).

Due to the presence of the solvation shell, electron transfer processes can happen differently. By adopting the terminology from coordination complex chemistry, the **outer-sphere** reaction is used to define the electron transfer processes where the solvation sphere is not negligible between the chemical species (such as an ion) and the electrode (Fig. 4.2b). In contrast, the **inner-sphere** reaction occurs when the chemical compound interacts strongly with the electrode (which are usually chemically specific), as demonstrated in Fig. 4.2c. Marcus theory describes the outer-sphere electron transfer processes as they can be treated in a more general way than the inner-sphere reactions.

In chemistry, the energy change of the chemical species during a chemical reaction is often represented in a graph where the x axis is an abstract one-dimensional **reaction coordinate** (q). Although the physical origin of the reaction coordinate is not clear, the phenomenological image of the energy path (inspired by experimental facts) is useful to understand the direction of chemical reactions. It is then important to obtain a general picture for redox reactions where the energy is a function of q.

Marcus imposed a harmonic oscillator assumption where the Gibbs free energy of the redox species is proportional to the square of the reaction coordinate q. This indicates that q is assigned

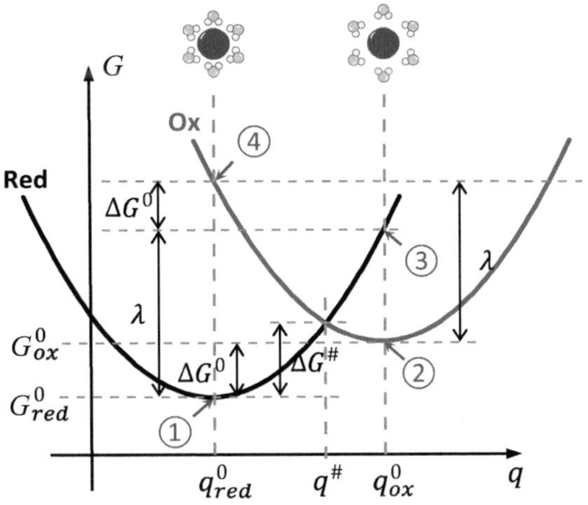

Figure 4.3 Upper panel: Different solvation configurations for chemical species at various locations along the reaction coordinates. Bottom panel: Gibbs free energy as a function of reaction coordinates.

with a physical meaning of the nuclear coordinate, and the nuclei fluctuation is based on the harmonic oscillator model. Figure 4.3 shows the parabolic energy profile of the redox species with respect to their nuclear coordinates. The Gibbs free energy of the redox species can be written as

$$G_{\text{red}} = \gamma(q - q_{\text{red}}^0)^2 + G_{\text{red}}^0, \tag{4.15a}$$

$$G_{\text{ox}} = \gamma(q - q_{\text{ox}}^0)^2 + G_{\text{ox}}^0 = \gamma(q - q_{\text{ox}}^0)^2 + \Delta G^0 + G_{\text{red}}^0, \tag{4.15b}$$

where γ represents the force constant (indicating the change in the bond length) and $\Delta G^0 = G_{\text{ox}}^0 - G_{\text{red}}^0$. q_{red}^0 and q_{ox}^0 are the equilibrium coordinates, and G_{red}^0 and G_{ox}^0 are the equilibrium free energy, of reduced and oxidized species, respectively (Fig. 4.3).

The solvation configuration of the chemical species varies along q, as illustrated in the upper panel in Fig. 4.3. States ① and ② indicate that the solvation configurations at q_{red}^0 and q_{ox}^0 are the most stable ones for the reduced and oxidized species, as G_{red}^0 and G_{ox}^0 are the local minima in the parabolic energy profiles. ③ and ④ show the unstable states for the reduced and oxidized species, as the solvation configuration gives a higher energy.

$\Delta G^{\#}$ in Fig. 4.3 represents the **activation energy** of the redox reaction. Note that $\Delta G^{\#}$ and λ can be expressed as

$$\Delta G^{\#} = \gamma(q^{\#} - q^0_{\text{red}})^2 = \Delta G^0 + \gamma(q^{\#} - q^0_{\text{ox}})^2, \tag{4.16}$$

$$\lambda = \gamma(q^0_{\text{ox}} - q^0_{\text{red}})^2. \tag{4.17}$$

It is then straightforward to obtain a relation

$$\sqrt{\Delta G^{\#}}(\sqrt{\Delta G^{\#}} - \sqrt{\lambda}) = \gamma(q^{\#} - q^0_{\text{red}})(q^{\#} - q^0_{\text{ox}}). \tag{4.18}$$

Hence,

$$\begin{aligned} \lambda &= \gamma(q^0_{\text{ox}} - q^0_{\text{red}})^2 \\ &= \gamma[(q^0_{\text{ox}} - q^{\#}) + (q^{\#} - q^0_{\text{red}})]^2 \\ &= 2\Delta G^{\#} - \Delta G^0 - 2(\Delta G^{\#} + \sqrt{\Delta G^{\#}\lambda}). \end{aligned} \tag{4.19}$$

By some mathematical rearrangement, the relation between $\Delta G^{\#}$ and λ can be written as

$$\Delta G^{\#} = \frac{(\Delta G^0 + \lambda)^2}{4\lambda}. \tag{4.20}$$

The reorganization energy (or the solvation energy, λ) is the energy difference between states ① and ③ (or ② and ④) in Fig. 4.3. It can be seen from Eq. 4.20 that λ plays an important role in electrochemistry analyses. Therefore many efforts have been invested to calculate the reorganization energy. Readers who are interested in this topic are recommended to consult [Bard and Faulkner (2000); Marcus (1956)].

The **electron transfer rate** can be calculated using the empirical **Arrhenius equation** [Bard and Faulkner (2000)],

$$k_{et} = \kappa \nu \exp\left(-\frac{\Delta G^{\#}}{k_B T}\right), \tag{4.21}$$

where κ is a transmission coefficient ranging between 0 and 1, and ν is the frequency of nuclear motion through the transition state ($\nu \sim \frac{k_B T}{h} \in (10^{12}, 10^{13})\ \text{s}^{-1}$). The product of κ and ν is defined as the frequency factor, and it can be predicted by statistical mechanics. The electron transfer rate for the homogeneous reactions has the dimension of s^{-1}, as can be inferred from Eq. 4.21. However, the unit of k_{et} changes to m/s (or cm/s, which is similar to that of velocity) for a heterogeneous system (such as the electrode–electrolyte system [Bard and Faulkner (2000)]), and then the electric current density can be obtained as

$$J = -enk_{et}, \tag{4.22}$$

where n is the number density of electrons being transfered.

4.1.3 The Gerischer Model

The Marcus theory described in the previous subsection is useful for calculating the electron transfer rate when different chemical species are analyzed, as the Gibbs free energy of the redox species are studied. As there was no discussion regarding the material properties of the electrode, it is difficult to describe the electrical processes at electrode–electrolyte interfaces using this theory by itself. It can be seen from Chapters 2 and 3 that the density of states of a material is useful to determine the electrical properties of a heterostructure by constructing the energy band diagram. For electrode–electrolyte systems, this then requires the development of a semiconductor-like theory for electrolyte, including the definition of equivalent density of states and Fermi level.

It should be noted that since the particles in electrolyte are classic and follow Maxwell–Boltzmann, instead of Fermi–Dirac, distribution, some definitions in semiconductor physics may not be applicable in solutions, as there are no quantum states in electrolyte. However, because electrons are transferred between electronic conductor (solid state electrode) and ionic conductor (electrolyte) to form current at the semiconductor–electrolyte interface, the correlation between work function and electrochemical potential (i.e., the equivalent Fermi level [Gerischer and Ekardt (1983)]) can still be used to describe the energy changes during electron transition processes in electrolyte.

Here we continue from the energy profile shown in Fig. 4.3, where states ①–④ are particularly of interest. As described previously, states ① and ② indicate the reduced and oxidized species are at their equilibrium state, while ③ and ④ are the states where the solvation shells of the reduced/oxidized species are the same as the oxidized/reduced species. These energy states are then reorganized in a fashion such that the energy of a free electron (in vacuum level) can be included, as demonstrated in Fig. 4.4. This is because the vacuum level of an electron is useful to determine its work function (energy difference between the vacuum and Fermi levels).

As illustrated in Fig. 4.4, the upper two levels have an accompanying electron while the lower two levels only have the

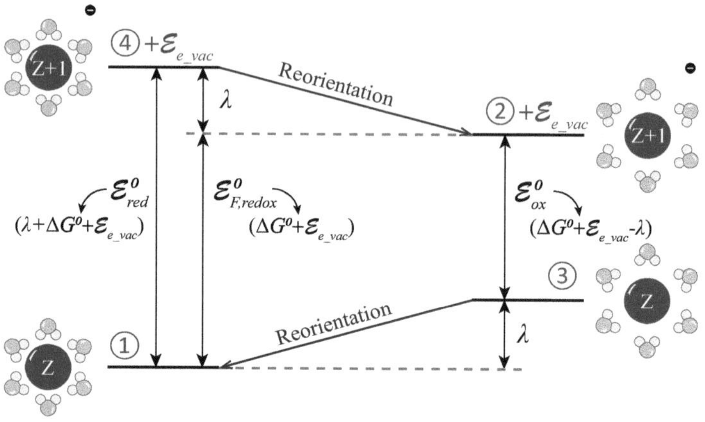

Figure 4.4 Energy differences between various redox species indicating the electron energy changes in electrolyte.

solvated ion. The number of charges for the reduced and oxidized species is Z and $Z+1$, respectively (Eq. 4.14). The energy differences between these energy levels are the electron energy in the redox system. Therefore, the *electron energy* ($\mathcal{E}$) can be written as

$$\begin{aligned} \mathcal{E} &= G_{\text{ox}} + \mathcal{E}_{\text{e\_vac}} - G_{\text{red}} \\ &= \gamma(q - q_{\text{ox}}^0)^2 - \gamma(q - q_{\text{red}}^0)^2 + \mathcal{E}_{\text{F,redox}}^0. \end{aligned} \tag{4.23}$$

where γ is a force constant, $\mathcal{E}_{\text{e\_vac}}$ is the energy of a free electron (at vacuum level), and $\mathcal{E}_{\text{F,redox}}^0 = G_{\text{ox}}^0 + \mathcal{E}_{\text{e\_vac}} - G_{\text{red}}^0 = \Delta G^0 + \mathcal{E}_{\text{e\_vac}}$ is the equivalent Fermi level for electrons in redox system. Since G_{red}^0 and G_{ox}^0 are the Gibbs energy of the solvated redox ions, and the difference between these ions is only that the reduced species contain more electrons, ΔG^0 represents the free energy of the electrons at equilibrium states. Therefore, $\Delta G^0 + \mathcal{E}_{\text{e\_vac}}$ is defined as the equivalent Fermi level (electrochemical potential), which is the same as that defined in Eq. 4.12.

Following the similar derivation of Eq. 4.20, we are able to obtain the relation between the energy of redox species and electrons after rearranging Eq. 4.23:

$$G_{\text{ox}} - G_{\text{ox}}^0 = \frac{(\mathcal{E} - \mathcal{E}_{\text{F,redox}}^0 + \lambda)^2}{4\lambda}, \tag{4.24a}$$

$$G_{\text{red}} - G_{\text{red}}^0 = \frac{(\mathcal{E} - \mathcal{E}_{\text{F,redox}}^0 - \lambda)^2}{4\lambda}. \tag{4.24b}$$

Solvated ions in solution follow Boltzmann distribution, and therefore the normalized density distribution for the oxidized and reduced species can be described by

$$W_{\text{ox}}(G_{\text{ox}}) = W_0 \exp\left(-\frac{G_{\text{ox}} - G_{\text{ox}}^0}{k_B T}\right), \tag{4.25a}$$

$$W_{\text{red}}(G_{\text{red}}) = W_0 \exp\left(-\frac{G_{\text{red}} - G_{\text{red}}^0}{k_B T}\right), \tag{4.25b}$$

where $W_0 = (4\pi\lambda k_B T)^{-\frac{1}{2}}$ is the normalization factor derived from $\int W(\mathcal{E})d\mathcal{E} = 1$. The unit for $W(\mathcal{E})$ is therefore $(\text{energy})^{-1}$. Note that Eq. 4.25 also indicates *the distribution of ions per energy, which are available for electrons to occupy.*

Substituting Eq. 4.24 into Eq. 4.25, and taking the concentration of the redox species into account, we are able to obtain the **density of (solvation) states** for electrons in redox system

$$\begin{aligned} D_{\text{ox}}(\mathcal{E}) &= c_{\text{ox}} W_{\text{ox}}(\mathcal{E}) \\ &= c_{\text{ox}} W_0 \exp\left\{-\frac{[\mathcal{E} - (\mathcal{E}_{\text{F,redox}}^0 - \lambda)]^2}{4\lambda k_B T}\right\}, \end{aligned} \tag{4.26a}$$

$$\begin{aligned} D_{\text{red}}(\mathcal{E}) &= c_{\text{red}} W_{\text{red}}(\mathcal{E}) \\ &= c_{\text{red}} W_0 \exp\left\{-\frac{[\mathcal{E} - (\mathcal{E}_{\text{F,redox}}^0 + \lambda)]^2}{4\lambda k_B T}\right\}, \end{aligned} \tag{4.26b}$$

where c_{ox} and c_{red} are the concentration for the oxidized and reduced species, respectively. It is then useful to define the electron energy levels in reduced and oxidized species ($\mathcal{E}_{\text{red}}^0$ and $\mathcal{E}_{\text{ox}}^0$) at equilibrium states by the equivalent Fermi level as

$$\mathcal{E}_{\text{ox}}^0 = \mathcal{E}_{\text{F,redox}}^0 - \lambda, \tag{4.27a}$$

$$\mathcal{E}_{\text{red}}^0 = \mathcal{E}_{\text{F,redox}}^0 + \lambda. \tag{4.27b}$$

Figure 4.5a illustrates the energy diagram of an electron which includes $\mathcal{E}_{\text{ox}}^0$, $\mathcal{E}_{\text{red}}^0$, and $\mathcal{E}_{\text{F,redox}}^0$ [Bard et al. (2002); Gellings and Bouwmeester (1997)]. It can be seen that the energy required for an electron to transit from the energy level of the reduced ($\mathcal{E}_{\text{red}}^0$) species to the energy of the oxidized ($\mathcal{E}_{\text{ox}}^0$) species is 2λ, where λ is the solvation energy. This is because an electron requires an

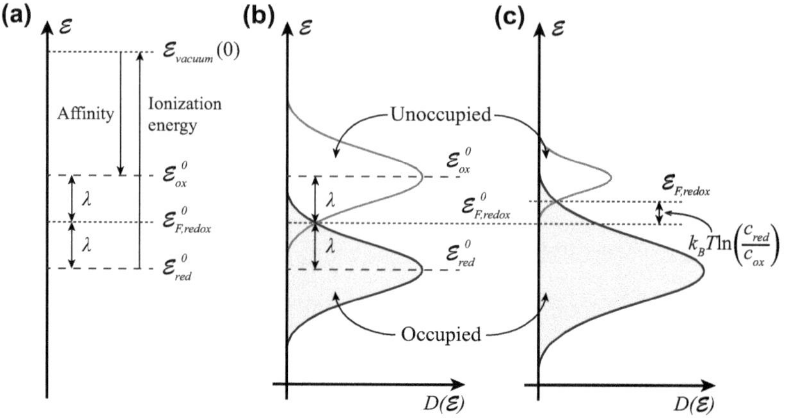

Figure 4.5 Electronic energy levels in redox system. (a) Simplified electron energy levels. (b-c) Electron energy level systems including the density distribution, where $c_{\text{ox}} = c_{\text{red}}$ in (b), and $c_{\text{ox}} < c_{\text{red}}$ in (c).

energy of 2λ to overcome the solvation shell in order to transfer from reduced to oxidized species, due to the ion solvation in electrolyte. $\mathcal{E}^0_{\text{F,redox}}$ represents the average energy level for 'free' electrons in a redox system. The energy difference between the vacuum level of the electrons ($\mathcal{E}_{\text{vacuum}}$) and $\mathcal{E}^0_{\text{ox}}$ is the **affinity**, and the difference between $\mathcal{E}_{\text{vacuum}}$ and $\mathcal{E}^0_{\text{red}}$ is the **ionization energy**. These are all concepts similar to those found in semiconductor physics and are very useful to study the electronic process at semiconductor–electrolyte interfaces. The origin and the relation to ionic energies of each of these terms in the electrolyte is explained in this section.

The most commonly used electronic energy system in electrolyte is illustratively demonstrated in Fig. 4.5b, where the electron density distributions (Eq. 4.26) are also included. If $c_{\text{ox}} = c_{\text{red}}$, the shapes of the density distribution for the reduced and oxidized species are identical. The Gaussian shape of the density distribution originates from the Maxwell–Boltzmann statistics. As electrons are transferred from the oxidized species to the reduced ones, the unoccupied and occupied sates for electrons in electrolyte are defined in Fig. 4.5b. It can be seen from Eq. 4.26 that $\mathcal{E} = \mathcal{E}^0_{\text{F,redox}}$ when $D_{\text{ox}}(\mathcal{E}) = D_{\text{red}}(\mathcal{E})$ and $c_{\text{ox}} = c_{\text{red}}$, which also gives the definition of $\mathcal{E}^0_{\text{F,redox}}$.

It is shown in Fig. 4.5c that the shape of the density distribution varies with the concentration of the redox species. This can be verified from Eq. 4.26. Note that the equivalent Fermi level in this case ($\mathcal{E}_{\mathrm{F,redox}}$) changes by $k_B T \ln\left(\frac{c_{\mathrm{red}}}{c_{\mathrm{ox}}}\right)$. This difference between $\mathcal{E}_{\mathrm{F,redox}}$ and $\mathcal{E}^0_{\mathrm{F,redox}}$ can be easily obtained by letting $D_{\mathrm{ox}}(\mathcal{E}_{\mathrm{F,redox}}) = D_{\mathrm{red}}(\mathcal{E}_{\mathrm{F,redox}})$.

The Gerischer model provides us with a powerful tool to study the electrical behavior at semiconductor–electrolyte interfaces. For example, the density of states and the Fermi level can be employed to calculate the electrical current during a redox reaction when different materials are used, and they are also useful for the analysis of the interfacial capacitance as well as photoelectrochemical processes.

4.2 Faradaic Processes

There are generally two types of processes occurring at electro-electrolyte interfaces. The first one involves the electron transfer between the electrode and electrolyte. This induces the redox reactions of the chemical species. As can be seen from Eq. 4.14, *the amount of the electrons being transferred is the same as the amount of the chemical reaction*, and this is the **Faraday's law** of electrolysis. Since the number of the electrons consumed is controlled by the flow of electricity between the electrode and electrolyte, the electrical current is often referred as the Faradaic current, and this process is the **Faradaic process**. In contrast, if there is no electron transfered, the process is **non-Faradaic**.

Investigation of Faradaic processes at electrode–electrolyte interfaces is enormously important for modern technology. This is because the potential applied across the electrode–electrolyte interface allows the control of the chemical reaction in the solution, or/and the phase transition of an element (between the chemical species in electrolyte and the solid material in electrode). This technique enables us to develop a wide range of applications. (i) As the redox potentials ($\mathcal{E}^0_{\mathrm{ox}}$ and $\mathcal{E}^0_{\mathrm{red}}$) usually vary with the type of molecules, molecular detection methods can then be

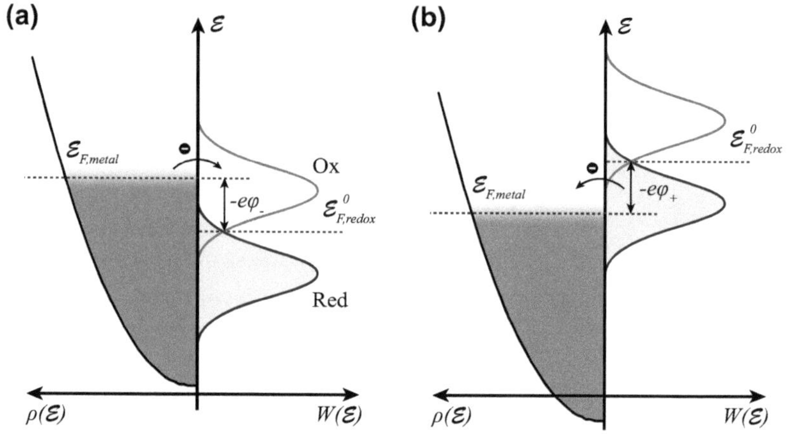

Figure 4.6 Fermi level alignment at a metal–electrolyte interface where the densities of states for the metal ($\rho(\mathcal{E})$) and the redox species ($W(\mathcal{E})$) are included. The applied potential is **(a)** negative (φ_-) and **(b)** positive (φ_+) on the metal with respect to the electrolyte.

developed by measuring $\mathcal{E}^0_{\text{ox}}$ and $\mathcal{E}^0_{\text{red}}$. (ii) If the products of the redox reactions stay on the surface of the electrode, we can use this technique to deposit thin films on the selected electrodes (electroplating). (iii) If the material of the electrode was involved in the redox reactions and then dissolves in the electrolyte, we are able to control the corrosion process using the electrochemical methods.

As can be seen from these examples, the Faradaic current plays a critical role in these applications. It is then essential to develop a method to describe this electrical current. Although Eq. 4.22 is adequate for the calculation of Faradaic current in some cases where the electrical properties of the electrode does not have to be taken into account, it is important to derive a general expression for all electrodes including semiconductors (where the electrical properties greatly differ from metals).

The Gerischer model gives a semiconductor-like energy level system for electrolyte (Fig. 4.5), this allows us to adopt the band alignment technique in section 2.3.2 to calculate the number (density) of the transferred electrons as well as the Faradaic current (density). Figure 4.6 demonstrates the energy levels at the metal–

electrolyte interface, where the densities of states for the metal ($\rho(\mathcal{E})$) and the redox species ($W(\mathcal{E})$) are illustrated. Note that $W(\mathcal{E})$ is the normalized density of states which represents the occupation probability of electrons in redox species.

It is seen from Fig. 4.6 that the applied potential across the metal–electrolyte interface is determined by the energy difference in the Fermi levels of the system, i.e., $-e\varphi = \mathcal{E}_{\mathrm{F,metal}} - \mathcal{E}^0_{\mathrm{F,redox}}$. When the applied potential is negative/positive on the metal with respect to the electrolyte, $\mathcal{E}_{\mathrm{F,metal}}$ is greater/smaller than $\mathcal{E}^0_{\mathrm{F,redox}}$. Hence, most electrons are transferred from the metal to the oxidized species in the electrolyte, and therefore reduce the chemical compounds. It is interesting to note that, regardless the chemical reaction in the solution, this electron transfer process is analogous to the quantum tunneling phenomenon in a scanning tunneling microscope where the electrons tunnel through a barrier between the sample and tip (usually made of different materials). The tunneling current is derived by calculating the tunneling rate which requires the density of states for both the sample and tip when using Fermi's golden rule [Grabert et al. (1992); Sakurai and Napolitano (2011)].

Similarly, the density of states of the electrode and electrolyte must be taken into account when deriving the Faradaic current. Although the number density of electrons in metal is given by $\int f(\mathcal{E})\rho(\mathcal{E})\mathrm{d}\mathcal{E}$, where $f(\mathcal{E})$ is the Fermi–Dirac distribution of the electrons, only a fraction of these electrons are useful to the reduction process at the metal–electrolyte interface (determined by the normalized distribution, $W_{\mathrm{ox}}(\mathcal{E})$). Therefore, the electrical current density contributed to the reduction is written

$$J_- = -ec_{\mathrm{ox}}\kappa\nu\sigma_{\mathrm{ox}}\sigma_{\mathrm{el}}\int_{-\infty}^{\infty} f(\mathcal{E})\rho(\mathcal{E})W_{\mathrm{ox}}(\mathcal{E})\mathrm{d}\mathcal{E}, \qquad (4.28)$$

where κ is the previously mentioned dimensionless transmission coefficient ranging between 0 and 1 (or defined as adiabadicity coefficient [Gerischer (1990)]), ν is the averaged frequency of nuclear motion through the activated complex, and σ_{ox} and σ_{el} represent the reaction cross-sections for the oxidized species and electrode, respectively.

It should be noted that, even for the case where $\mathcal{E}_{\mathrm{F,metal}}$ is greater than $\mathcal{E}^0_{\mathrm{F,redox}}$ (Fig. 4.6a), it is still possible for the electrons to transmit from the solution to electrode. It is more convenient to analyze this process by envisaging holes transferring from the unoccupied band in the electrode to the reduced species in the electrolyte (which is filled by electrons, and vacant for holes), due to the factors including thermal excitation and so on. In this case, the electrical current density for the electrons flowing backwards from the electrolyte to electrode is written

$$J_+ = -ec_{\text{red}}\kappa\nu\sigma_{\text{red}}\sigma_{\text{el}}\int_{-\infty}^{\infty} f(\mathcal{E})\rho(\mathcal{E})W_{\text{red}}(\mathcal{E})\text{d}\mathcal{E}, \tag{4.29}$$

where the subscript *red* denote that the physical quantity is defined for the reduced species in the solution. Therefore, the net electrical current density is

$$J = J_- - J_+. \tag{4.30}$$

This is the Faradaic current density at metal–electrolyte interfaces. Note that when the applied potential on the metal is positive with respect to electrolyte (Fig. 4.6b), the Faradaic current density is still calculated using Eqs. 4.28–4.30, except for that the direction of the net current will be opposite to that for Fig. 4.6a. This is because no assumptions or approximations related to the position of the Fermi levels are used in the derivation process for these equations.

At semiconductor–electrolyte interfaces, where there is a bandgap for the semiconductor, the integration limits in Eqs. 4.28 and 4.29 must be changed. Hence, the electrical current density for the reduction and oxidation processes are

$$\begin{aligned} J_- = &-ec_{\text{ox}}\nu\sigma_{\text{ox}}\left[(\kappa\sigma_{\text{el}})_{\mathcal{E}_C}\int_{\mathcal{E}_C}^{\infty} f(\mathcal{E})\rho(\mathcal{E})W_{\text{ox}}(\mathcal{E})\text{d}\mathcal{E}\right. \\ &\left.+ (\kappa\sigma_{\text{el}})_{\mathcal{E}_V}\int_{-\infty}^{\mathcal{E}_V} f(\mathcal{E})\rho(\mathcal{E})W_{\text{ox}}(\mathcal{E})\text{d}\mathcal{E}\right], \end{aligned} \tag{4.31a}$$

$$\begin{aligned} J_+ = &-ec_{\text{red}}\nu\sigma_{\text{red}}\left[(\kappa\sigma_{\text{el}})_{\mathcal{E}_C}\int_{\mathcal{E}_C}^{\infty} f(\mathcal{E})\rho(\mathcal{E})W_{\text{red}}(\mathcal{E})\text{d}\mathcal{E}\right. \\ &\left.+ (\kappa\sigma_{\text{el}})_{\mathcal{E}_V}\int_{-\infty}^{\mathcal{E}_V} f(\mathcal{E})\rho(\mathcal{E})W_{\text{red}}(\mathcal{E})\text{d}\mathcal{E}\right], \end{aligned} \tag{4.31b}$$

where the subscripts $\mathcal{E}_C$ and $\mathcal{E}_V$ for $(\kappa\sigma_{\text{el}})$ indicate that the value of $(\kappa\sigma_{\text{el}})$ is different at the conduction and valence band edges. The expression of the net Faradaic current density is still the same as Eq. 4.30. However, it should be noted that, as graphene is a semimetal with no bandgaps, Eqs. 4.28–4.30 are adequate for graphene–electrolyte systems.

4.3 Non-Faradaic Processes

When an electrode is in contact with an electrolyte, redox reactions do not always take place at the interface. In the cases where no charge transfer occur at the electrode–electrolyte interface, we regard these processes as non-Faradaic and define the electrodes as ideal polarized electrodes (IPE). Although there is no perfect IPE in reality, it can be inferred from Fig. 4.6 that the Faradaic current can be kept at a minimum level (which is negligible in some cases) if the potential on the electrode remains in a very small range near $\mathcal{E}^0_{\text{F,redox}}$. This process is particularly useful for the capacitive applications, because the charges stay at the electrode–electrolyte interface without inducing charge transfer current (same as the definition of a capacitor). It is then of interest to investigate how the charges are stored at the interface, i.e., the ionic structure.

4.3.1 Gouy–Chapman–Stern Theory

In this case, an interesting structure composed of solvated ions is formed at the interface of electrode–electrolyte systems. The formation of this structure has been conventionally studied using a theory proposed by G. Gouy and D. L. Chapman in 1910s [Chapman (1913); Gouy (1917, 1910)] and then further developed by O. Stern in 1924 [Stern (1924)]. In the Gouy–Chapman–Stern (GCS) theory, two layers of solvated ions arranged differently at the electrode–electrolyte interface, and form the structure of **electrical double layer** (EDL). The EDL is a critical component for a wide range of applications, such as supercapacitors, fuel cells, implantable neuromodulation devices and more [Simon and Gogotsi (2008); Zhan et al. (2016)], it is therefore of interest to investigate the EDL structure, as well as the resulting electrochemical capacitance.

4.3.1.1 The Gouy–Chapman theory

Ions in electrolyte are classical, and they follow the Maxwell–Boltzmann statistics (Eq. 2.11). This is because the classical identical particles can be distinguished from each other (by tracing their trajectories), and the Pauli exclusion principle does not apply.

As can be seen from the dashed curves in Fig. 2.3b–d that the Maxwell–Boltzmann distribution at a nontrivial energy (e.g., greater than 0.2 eV) simply follows a logarithm law. Therefore, when an electrostatic energy is present in an electrolyte due to the applied electrical potential, the ion density at a specific location can be written as

$$n_i = n_i^0 \exp\left[-\frac{z_i e(\varphi - \varphi^0)}{k_B T}\right], \tag{4.32}$$

where the subscript i represents the species of ions (i.e., cations or anions). z is the signed charge, e the charge on the electron, k_B the Boltzmann constant, and T the absolute temperature. While φ is the electrical potential at this location, n_i^0 and φ^0 are the ion density and the electrical potential at a reference point (usually the bulk solution). Due to the quasi-electroneutrality in the bulk electrolyte [Bard and Faulkner (2000)], we often have $\varphi^0 = 0$.

When an electrode with a non-zero potential is immersed in a liquid, dissolved ions with opposite charges are attracted to the surface of the electrode. The Gouy–Chapman theory describes a distribution of the ions at a charged surface (diffuse layer) as sizeless point charges using Maxwell–Boltzmann statistics, as shown in Fig. 4.7a. When studying a practical problem, it is useful to find

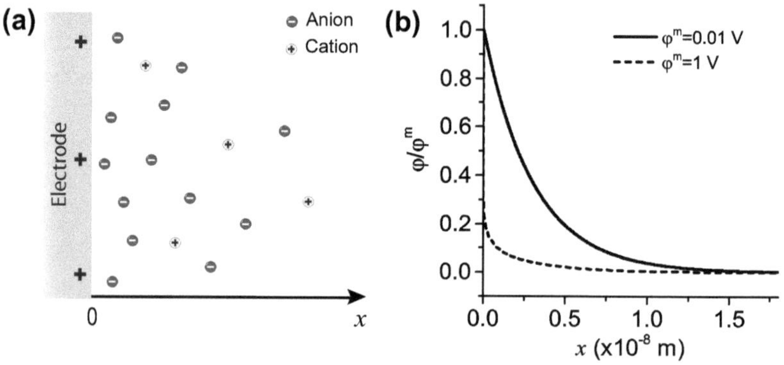

Figure 4.7 (a) The Gouy–Chapman model of ions at an electrode–electrolyte interface, which are considered as sizeless point charges following the Maxwell–Boltzmann distribution. (b) The ratio of the potential in electrolyte (φ) and that applied on the electrode (φ^m) as a function of position (x).

the electrical potential distribution in electrolyte, particularly at the electrode–electrolyte interface. Substituting Eq. 4.32 into the Poisson's equation gives the **Poisson–Boltzmann equation**

$$\frac{\mathrm{d}^2\varphi}{\mathrm{d}x^2} = -\frac{e}{\epsilon}\sum_i z_i n_i^0 \exp\left(-\frac{z_i e\varphi}{k_\mathrm{B}T}\right), \tag{4.33}$$

where ϵ is the absolute electric permittivity (dielectric constant). x represents the distance from the electrode–electrolyte interface to the bulk solution. The sum over all the cations and anions including the signed charge z_i indicates that only the net charge is considered. It is more convenient to rewrite Eq. 4.33 to

$$\left(\frac{\mathrm{d}\varphi}{\mathrm{d}x}\right)^2 = \frac{2k_\mathrm{B}T}{\epsilon}\sum_i n_i^0 \left[\exp\left(-\frac{z_i e\varphi}{k_\mathrm{B}T}\right) - 1\right], \tag{4.34}$$

where a relation $\frac{\mathrm{d}^2\varphi}{\mathrm{d}x^2} = \frac{1}{2}\frac{\mathrm{d}}{\mathrm{d}\varphi}\left(\frac{\mathrm{d}\varphi}{\mathrm{d}x}\right)^2$ is used, and a boundary condition $\left(\left.\frac{\mathrm{d}\varphi}{\mathrm{d}x}\right|_{\varphi=\varphi^0=0} = 0\right)$ is applied.

For symmetric electrolyte in which the absolute value of the charge for the cations and anions is the same (i.e., $z_c = -z_a = z$ where the subscripts c and a denote cations and anions, respectively), the solution of Eq. 4.34 can be obtained,

$$\frac{\tanh\left(\frac{ze\varphi}{4k_\mathrm{B}T}\right)}{\tanh\left(\frac{ze\varphi^m}{4k_\mathrm{B}T}\right)} = \exp\left(-\frac{x}{\lambda_\mathrm{D}}\right), \tag{4.35}$$

where φ^m represents the potential applied on the electrode surface ($x = 0$). $\lambda_\mathrm{D} = \sqrt{\frac{\epsilon k_\mathrm{B}T}{2n^0(ze)^2}}$ is the Debye length, and $n^0 = n_i^0$. Note that there is a difference of 2 in the denominator inside the square root compared with Eq. 4.13. This comes from the sum over the cations and anions. Equation 4.35 indicates that the electric potential decays rapidly from the electrode–electrolyte interface into the solution. Figure 4.7b shows the variation of electrical potential as a function of x for a symmetric electrolyte where the concentration is 1 M (i.e., 1 mole/L) and $z = 1$. In spite of the hyperbolic tangent functions included in Eq. 4.35, the potential in electrolyte decreases exponentially when φ^m is small enough (e.g., $\varphi^m = 0.01$ V), as demonstrated by the solid curve (which overlaps with an exponential relation) in Fig. 4.7b. It can be also seen that when φ^m increases, the decaying rate of φ becomes faster.

It is then important to calculate the number of charges stored at the electrode–electrolyte interface. The Neumann boundary condition for the Poisson's equation states that the surface charge density at an interface of two media is equal to the difference between the normal components of the electric displacement field (D) at each side of the interface [Jackson (2007)]. For the case shown in Fig. 4.7(a), $D = -\epsilon\frac{\partial\varphi}{\partial x}$, where $-\frac{\partial\varphi}{\partial x}$ represents the x component of the electric field. Since the electric field inside the metal $\left(-\frac{\partial\varphi}{\partial x}\right)^m$ is 0, the net surface-charge density on the metal is obtained

$$\begin{aligned}\sigma^m &= \left(-\epsilon\frac{\partial\varphi}{\partial x}\right)^e_{x=0} - \left(-\epsilon\frac{\partial\varphi}{\partial x}\right)^m_{x=0} \\ &= \sqrt{8k_\mathrm{B}T\epsilon n^0}\sinh\left(\frac{ze\varphi^m}{2k_\mathrm{B}T}\right),\end{aligned} \tag{4.36}$$

where the superscripts e and m denote electrolyte and metal, respectively. Note that the square root of Eq. 4.34 is used to represent $\left(\frac{\partial\varphi}{\partial x}\right)^e$. Since the Neumann boundary condition is a direct result from the Gauss' law, same results can be obtained by constructing a Gauss box in the metal with one side approaching arbitrarily close to the metal–electrolyte interface.

Gauss boxes can also be constructed in many other ways to serve different purposes. For example, if a Gauss box is selected as that shown in Fig. 4.8a, it can be seen that the number of charges accumulated on the electrode surface is the same as that distributed in the electrolyte. This is because the electric field at each side of the box is 0 (due to the electroneutrality inside the metal and the bulk electrolyte). Therefore, $\sigma^m = -\sigma^e$, where σ^e is the stored net charges per area (the normal direction of the area is along the x axis). σ^e can be derived by integrating the volumetric net charge density in electrolyte over x:

$$\sigma^e = \int_0^{\mathrm{bulk}} \rho_{\mathrm{net}}\mathrm{d}x, \tag{4.37}$$

where ρ_{net} can be derived from Eq. 4.32,

$$\begin{aligned}\rho_{\mathrm{net}} &= zen^0\left[\exp\left(-\frac{ze\varphi}{k_\mathrm{B}T}\right) - \exp\left(\frac{ze\varphi}{k_\mathrm{B}T}\right)\right] \\ &= -2zen^0\sinh\left(\frac{ze\varphi}{k_\mathrm{B}T}\right).\end{aligned} \tag{4.38}$$

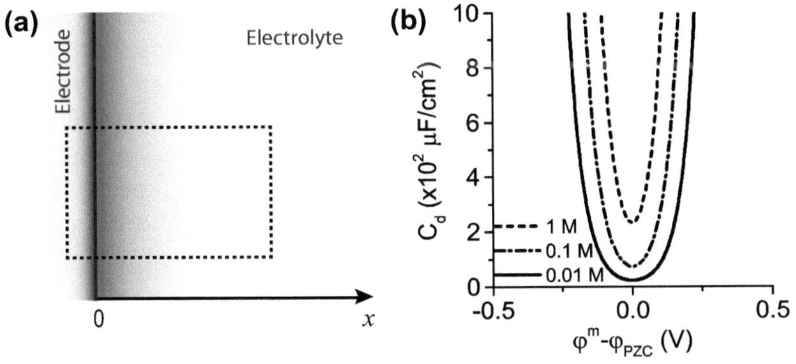

Figure 4.8 (a) Selection of a Gauss box (dashed lines) across the electrode–electrolyte interface, where the charge density distribution is illustrated by the gray scale along x axis. (b) The calculated differential capacitance as a function of electrode voltage (with respect to the potential of zero charge φ_{PZC}) for a symmetric electrolyte ($z = 1$) of different concentrations.

These relations are useful to calculate σ^m (or σ^e) of the EDL for applications such as supercapacitors or non-Faradaic biosensors. It is worth mentioning that at a certain potential, there is no charge stored ($\sigma^m = -\sigma^e = 0$). This is defined as the **potential of zero charge** (φ_{PZC}).

As described earlier, one of the most common applications of non-Faradaic process at electrode–electrolyte interface is to develop capacitors. It is then important to derive the areal differential capacitance (C_d) of the system, which is defined as the derivative of the areal density of the stored net charge (σ^m) with respect to the applied potential (φ^m)

$$C_\text{d} = \frac{\text{d}\sigma^m}{\text{d}\varphi^m} = \frac{\epsilon}{\lambda_\text{D}} \cosh\left(\frac{ze\varphi^m}{2k_\text{B}T}\right), \tag{4.39}$$

where λ_D is defined in Eq. 4.35. Note that C_d determines the rate that the stored charge density changes with the applied potential, in contrast to the algebraic capacitance ($\frac{\sigma^m}{\varphi^m}$) which simply defines the stored charges at a certain potential.

Figure 4.8b illustrates the calculated differential capacitance as a function of electrode voltage (with respect to φ_{PZC}) for a symmetric electrolyte ($z = 1$) of various concentrations. It can be seen that C_d increases with the concentration of electrolyte. It is worth noting that there is a minimum in C_d at the point of $\varphi^m - \varphi_{PZC} = 0$, in spite of different concentrations. This is because, when the electrode is neutral ($\varphi^m = \varphi_{PZC}$), ions are not attracted or repelled by the surface resulting in surface ion concentration, which follows an exponential relation (Eq. 4.32), equivalent to that of the bulk solution and thus a capacitance minimum. This V-shaped variance of capacitance at moderate electrode voltages, around the φ_{PZC}, is confirmed experimentally.

It can also be seen from Fig. 4.8b and Eq. 4.39 that C_d could increase to infinity if φ^m keeps increasing. This, however, is not realistic and is not observed in experiments. This is because the Gouy–Chapman model allows an ion to approach arbitrarily close to the electrode as a sizeless point charge at non-zero voltages. This indicates that a further development of this theory is necessary.

4.3.1.2 The Stern layer

In reality, ions have finite size and they cannot approach arbitrarily close to the electrode. This has to be taken into account when deriving the potential distribution and the differential capacitance. Based on this idea, Stern introduced a layer of solvated ions with fixed thickness in the close proximity of the electrode, as illustrated in Fig. 4.9a. This layer is called **Stern layer** (or outer Helmholtz plane, OHP), and it comprises the **electrical double layer** (EDL) with the **diffuse layer**, which is defined as the region between OHP and the bulk solution. Note that the inner Helmoholtz plane (IHP) is defined by the location of the specifically adsorbed ions in EDL [Bard and Faulkner (2000)], and it is not discussed here.

The thickness of Stern layer is then determined by the radius of solvated ions (r_H). Figure 4.9a shows the structure of an EDL, where x_2^0 denotes the position of the OHP and x_1^0 is simply equal to $x_2^0 + 2r_H$. As there is no net charge accumulated inside the Stern layer (between 0 and x_2^0), the electrical potential demonstrates a linear relation with x, i.e., the solvation shell is considered as a simple dielectric. Note that the introduction of Stern layer does not affect the ion distribution (Boltzmann statistics) in the diffuse layer. This suggests that the Poisson–Boltzmann equation and its solution (Eqs. 4.33 and 4.35) are still valid to describe the potential profile in the

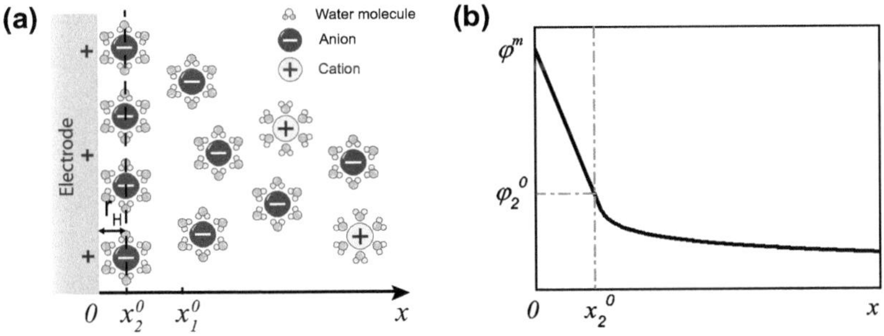

Figure 4.9 (a) Gouy–Chapman–Stern (GCS) model for the EDL at an electrode–electrolyte interface, where r_{H} represents the distance from the electrode surface to OHP (located at x_2^0). (b) Potential profile adjacent to an electrode surface in electrolyte, where φ^m is the electrode potential and φ_2^0 is the local potential at x_2^0.

diffuse layer

$$\frac{\tanh\left(\frac{ze\varphi}{4k_{\mathrm{B}}T}\right)}{\tanh\left(\frac{ze\varphi_2^0}{4k_{\mathrm{B}}T}\right)} = \exp\left(-\frac{x - x_2^0}{\lambda_{\mathrm{D}}}\right), \tag{4.40}$$

where φ_2^0 represents the electrical potential at x_2^0. It is then useful to write down the slope of φ at x_2^0,

$$\left(\frac{\mathrm{d}\varphi}{\mathrm{d}x}\right)_{x=x_2^0} = -\sqrt{\frac{8k_{\mathrm{B}}Tn^0}{\epsilon}}\sinh\left(\frac{ze\varphi_2^0}{2k_{\mathrm{B}}T}\right). \tag{4.41}$$

Note that this equation can be derived directly from Eq. 4.34.

Due to the Neumann and Dirichlet boundary conditions for the Poisson's equation, the electric field at x_2^0 is continuous. This gives a relation to find φ_2^0

$$\frac{\varphi_2^0 - \varphi^m}{x_2^0} = -\sqrt{\frac{8k_{\mathrm{B}}Tn^0}{\epsilon}}\sinh\left(\frac{ze\varphi_2^0}{2k_{\mathrm{B}}T}\right). \tag{4.42}$$

The solution of this equation can be derived numerically by the Newton–Raphson method [Burden and Faires (2010)]. Hence the electrical potential profile at the electrode–electrolyte interface is obtained. The solid curves in Fig. 4.10a show the calculated potential distribution at an electrode–electrolyte interface when the voltage applied on the electrode (φ^m) is 1 V and 0.3 V.

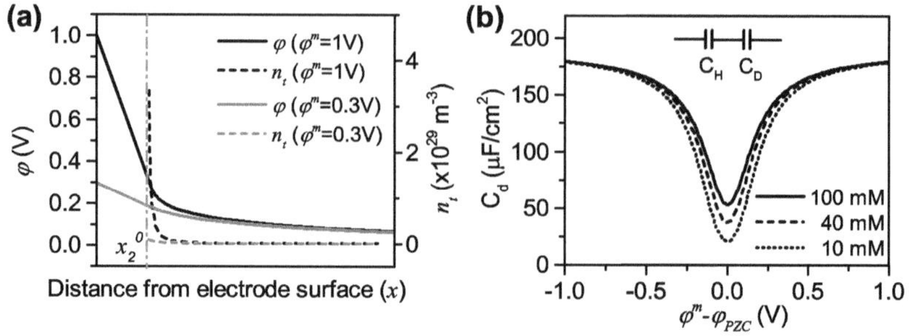

Figure 4.10 (a) Potential distribution (solid curves) at an electrode–electrolyte interface when the voltage applied on the electrode (φ^m) is 1 and 0.3 V, respectively. φ is the local potential at position x in solution. n_t (dashed curves) is the total volumetric density of both cations and anions. (b) Differential capacitance of a KF aqueous solution at different concentration, calculated by GCS theory. The equivalent circuit model for an ideal electrical double layer capacitor is shown in the inset, where C_{H} and C_{D} represent the equivalent capacitance for the Stern layer and the diffuse layer, respectively.

The surface density of the stored charges on the electrode is then written

$$\sigma^m = -\epsilon\left(\frac{\mathrm{d}\varphi}{\mathrm{d}x}\right)_{x=x_2^0} = \sqrt{8k_{\mathrm{B}}T\epsilon n^0}\sinh\left(\frac{ze\varphi_2^0}{2k_{\mathrm{B}}T}\right). \tag{4.43}$$

It can be seen that this equation has the same form as that of Eq. 4.36, except that φ^m in Eq. 4.36 is replaced by φ_2^0 here. As can be seen from Fig. 4.10a, φ_2^0 is usually smaller than φ^m, this suggests that the non-zero thickness of the Stern layer (i.e., the radius of solvated ions) reduces the number of charges stored on the electrode.

It is useful to write the inverse of the differential capacitance (C_{d})

$$\frac{1}{C_{\mathrm{d}}} = \left(\frac{\mathrm{d}\sigma^m}{\mathrm{d}\varphi^m}\right)^{-1} = \frac{x_2^0}{\epsilon} + \left[\sqrt{\frac{2(ze)^2\epsilon n^0}{k_{\mathrm{B}}T}}\cosh\left(\frac{ze\varphi^m}{2k_{\mathrm{B}}T}\right)\right]^{-1}$$
$$= \frac{1}{C_{\mathrm{H}}} + \frac{1}{C_{\mathrm{D}}}, \tag{4.44}$$

where C_{H} and C_{D} represent the differential capacitance of the Stern layer and the diffuse layer, respectively. This indicates that an EDL capacitor can be viewed as circuit consists of two capacitors (Stern layer and diffuse layer) in series. The circuit model is demonstrated in the inset of Fig. 4.10b.

The introduction of Stern layer in EDL prevents the differential capacitance from increasing to infinity, which was predicted by the Gouy–Chapman model. It provides a saturation level for C_d at high electrode potentials, as can be seen from Fig. 4.10b. This is because the capacitance contributed from the Stern layer becomes dominant in the serial circuit (as C_D would rise to a very large value when φ_2^0 increases, and the smaller capacitance dominates in such a circuit) and the thickness is fixed. The GCS theory delivers a clear physical picture for the structure of ions at the electrode–electrolyte interface, and it enables us to predict the differential capacitance of electrochemical systems. Therefore, this model is widely used in the scientific and engineering communities to obtain systematic understanding for non-Faradaic processes.

4.3.2 Modified Poisson–Boltzmann Model

In spite of the success of the GCS model in electrochemistry, a revisit to Eq. 4.32, however, indicates that the total ion density in the Stern layer could rise to a unrealistic value when the electrical potential on the electrode is high enough. The total volumetric density for ions which follow the Maxwell–Boltzmann statistics is

$$\begin{aligned} n_t &= n^0\left[\exp\left(-\frac{ze\varphi}{k_\mathrm{B}T}\right)+\exp\left(\frac{ze\varphi}{k_\mathrm{B}T}\right)\right] \\ &= 2n^0\cosh\left(\frac{ze\varphi}{k_\mathrm{B}T}\right). \end{aligned} \tag{4.45}$$

Integration of n_t from x_2^0 to x_1^0 (as indicated in Fig. 4.9a) gives the ionic areal density at the OHP located at x_2^0, $n_\mathrm{OHP}^\mathrm{2D} = \int_{x_2^0}^{x_1^0} n_t \mathrm{d}x$, where $x_1^0 = x_2^0 + 2r_\mathrm{H}$. $n_\mathrm{max}^\mathrm{2D}$ is defined as the maximal possible local areal density for all ions and molecules in a single ionic/molecular layer. Similar to solid state crystal structures, $n_\mathrm{max}^\mathrm{2D}$ can be obtained by arranging ions/molecules in a 2D version of a hexagonal or a centered rectangular lattice, hence $n_\mathrm{max}^\mathrm{2D} = (2\sqrt{3}r_\mathrm{H}^2)^{-1}$. At a sufficiently high potential, $n_\mathrm{OHP}^\mathrm{2D}$ might exceed $n_\mathrm{max}^\mathrm{2D}$, as demonstrated in Fig. 4.10a. For $\varphi_s = 1V$ in a 10 mM KF electrolyte, a steep increase of n_t is observed near the OHP, resulting in $n_\mathrm{OHP}^\mathrm{2D} = 8.4 \times 10^{18}m^{-2}$ (while $n_\mathrm{max}^\mathrm{2D} = 2.1 \times 10^{18}m^{-2}$). It suggests that the ionic density is greater than the limit. This, however, is not realistic, and the fact that

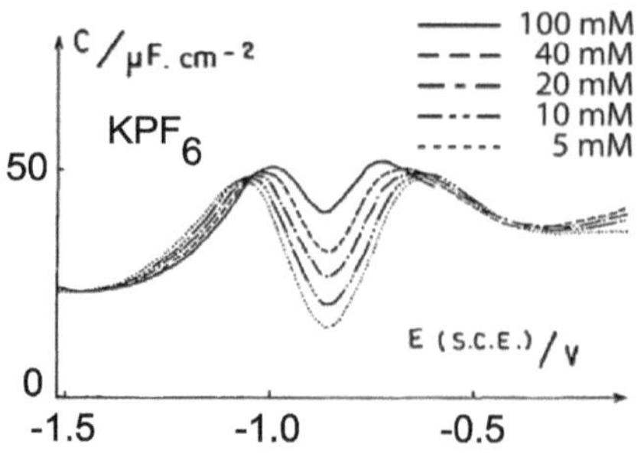

Figure 4.11 Measured differential capacitance of KPF_6 aqueous solution by silver electrode at different concentration. Reprinted with permission from [Valette (1982)]. Copyright Elsevier.

only a certain number of ions can fit onto the Stern layer is called the **steric effect**.

In addition, experiments show that, rather than saturating, the EDL capacitance is regularly observed to increase to a maximum and then decrease as electrode potentials move further away from φ_{PZC} [Islam et al. (2008); Lockett et al. (2008); Valette (1982)] (Fig. 4.11). The observed decrease of capacitance at high potentials is not predicted by GCS resulting in a large overestimation of electrode capacitance at high voltage. This suggests that the assumption of fixed thickness of the Stern layer may not be correct. The conventional Stern layer has its thickness fixed for only one layer of ions at the electrode surface. This means there is a limit for the areal density of ions on the electrode surface. However, when the actual areal density exceeds the limit created when forming a continuous monolayer as described previously, the surplus ions will have to stay on top of the initial Stern layer, and hence the effective "Stern layer" thickness increases, because only a certain numbers of ions can fit onto the surface. However, the GCS does not take this into account.

It is therefore necessary to understand how the surplus ions will "stack" on the original Stern layer. There are two major ways to tackle this problem. From the perspective of thermodynamics, the free energy of a system at equilibrium must stay minimum. It is then useful to derive the free energy of the electrochemical system, and perform a derivative to find the new differential capacitance. From the perspective of ion dynamics, the motion of ions in EDL must satisfy a series of fluid mechanical equations. While the first method is ideal for the study of a system at equilibrium, the second one is very useful to investigate the time-dependent behavior of ion dynamics in the EDL which does not necessarily require the system to stay at equilibrium. This subsection will briefly introduce the efforts made in the past to tackle the steric effect, and the time-dependent study of the ions in EDL will be presented in the next subsection.

Not long after J. J. Bikerman realized that the volume of ions needs to be taken into account when studying the differential capacitance in 1942 [Bikerman (1942)], M. Eigen and E. Wick calculated the free energy of the thermodynamical systems for electrolytes at high concentration in 1954 [Eigen and Wick (1954)]. The advances of this theory in the recent decades deliver a modified Boltzmann distribution function [Bazant et al. (2011); Borukhov et al. (1997); Kornyshev (2007); Stafiej et al. (1996)]. This leads to a deeper understanding of the electrical double layer at equilibrium, and the theory which accounts for the steric effect is known as the modified Poisson–Boltzmann model.

The total free energy of a system (F) measures the high-quality work (which can be converted into other forms of energy) of a thermodynamic system, and it is expressed as

$$F = U - TS - \mu^p n, \tag{4.46}$$

where U is the internal energy of the system, S is the entropy, and μ^p is the chemical potential. The internal energy contains the electrostatic contribution in the system, and it is written as

$$U = \int \mathrm{d}r \left(-\frac{\epsilon}{2} |\nabla \varphi|^2 + z_a e \varphi_a + z_c e \varphi_c \right), \tag{4.47}$$

where the first term is the field energy, and the next two terms represent the electrostatic energy of the ions. Note that the standard units (SI) are used. The steric effect is considered using a lattice gas model, in which each cell of the lattice can be occupied by at most one ion [Borukhov et al. (2000)]. This model is then used to derive the entropy of the system,

$$\begin{aligned} -TS = \frac{k_\mathrm{B} T}{a^3} \int \mathrm{d}r \, \big[& n_c a^3 \ln(n_c a^3) + n_a a^3 \ln(n_a a^3) \\ & + (1 - n_c a^3 - n_a a^3) \ln(1 - n_c a^3 - n_a a^3) \big] . \end{aligned} \tag{4.48}$$

where a represents the diameter of the ions, and the Stirling formula has been used [Kornyshev (2007)]. For simplicity, the size of anions and cations are considered the same. The last term in Eq. 4.46 can be simply written as

$$-\mu^p n = -\mu_c^p n_c - \mu_a^p n_a. \tag{4.49}$$

By minimizing the free energy of the system (finding the derivative of F with respect to φ and n), we obtain the ion density at the electrode–electrolyte interface:

$$n_c = n^0 \frac{\exp\left(-\frac{z_c e\varphi}{k_B T}\right)}{1 - \gamma + \gamma \cosh\left(\frac{z_c e\varphi}{k_B T}\right)}, \tag{4.50a}$$

$$n_a = n^0 \frac{\exp\left(-\frac{z_a e\varphi}{k_B T}\right)}{1 - \gamma + \gamma \cosh\left(\frac{z_a e\varphi}{k_B T}\right)}, \tag{4.50b}$$

where $\gamma = 2\frac{n^0}{n^{\max}} = 2a^3 n^0$ is a dimensionless constant accounting for the steric effect, and $n^{\max} = a^{-3}$ is the maximal possible local density of ions. The factor of 2 comes from the contribution of both cations and anions, and $0 \leq \gamma \leq 1$. It can be seen that Eq. 4.50 returns to Eq. 4.32 when $\gamma = 0$. If $\gamma > 0$, Eq. 4.50 describes as a Fermi-like distribution for the ions, other than the regular exponential Boltzmann statistics, at the electrode–electrolyte interface. $\gamma = 1$ represents the closed-pack condition where the ions are allocated compactly as those in solid crystals.

Equation 4.50 can be substituted into the Poisson's equation to calculate the electric potential profile, as well as the spatial distribution of ion density at the interface. This gives the modified Poisson–Boltzmann (mPB) equation

$$\nabla^2 \varphi = \frac{zen^0}{\epsilon} \frac{\sinh\left(\frac{ze\varphi}{k_B T}\right)}{1 - \gamma + \gamma \cosh\left(\frac{ze\varphi}{k_B T}\right)}, \tag{4.51}$$

where $z = z_c = -z_a$. Note that this equation is derived for the symmetric electrolyte for simplicity, and it can be solved numerically.

Figure 4.12 shows the calculated potential distribution at the electrode–electrolyte interface when γ and the voltages applied on

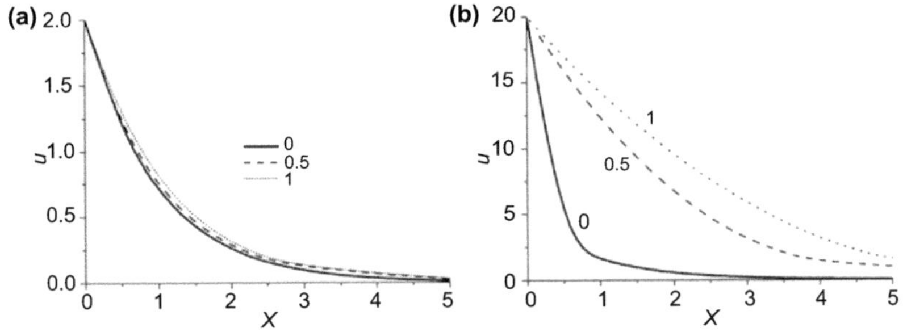

Figure 4.12 Electric potential profile at the electrode–electrolyte interface when $\gamma = 0, 0.5$, and 1. Note that the voltage applied on the electrode is (a) 50 mV (i.e., $u = 2$) and (b) 0.5 V (i.e., $u = 20$), respectively. Reprinted with permission from [Kornyshev (2007)]. Copyright American Chemical Society.

the electrode are different, where $u = \frac{ze\varphi}{k_B T}$ (e.g., $u = 39.6\varphi$ if $z = 1$). It can be seen from Fig. 4.12a that the potential distributions for $\gamma = 0$, 0.5, and 1 are very similar when the applied voltage on the electrode is low ($u = 2$, i.e., $\varphi = 50$ mV). This indicates that the steric effect does not affect the potential profile at the interface at low voltages. However, when the electrode voltage increases, for example, to $u = 20$ (i.e., $\varphi = 0.5$ V) in Fig. 4.12b, the potential in EDL decays at a much slower rate when $\gamma = 1$ (the closed-pack condition), in comparison to that for $\gamma = 0$ (the regular Maxwell–Boltzmann distribution). This indicates that the steric effect must be taken into account at high electrode potentials.

The variation of electric potential profile at high electrode voltages suggests that the ion density distribution in EDL should change as well. Figure 4.13a shows the calculated ion density at the interface using Eq. 4.50 and the solution of Eq. 4.51 for the different diameters of the ions (i.e., $a = 7.5$ Å and $a = 10$ Å). It can be seen that the distribution function has a Fermi-like shape when the steric effect is taken into account, because only certain number of ions per area can fit onto the electrode surface. Note that the regular Boltzmann relation is included in the figure for comparison. The drastic change between the ion density distributions with and without the steric effects suggests that the differential capacitance of the EDL may be different from the prediction of the GCS model.

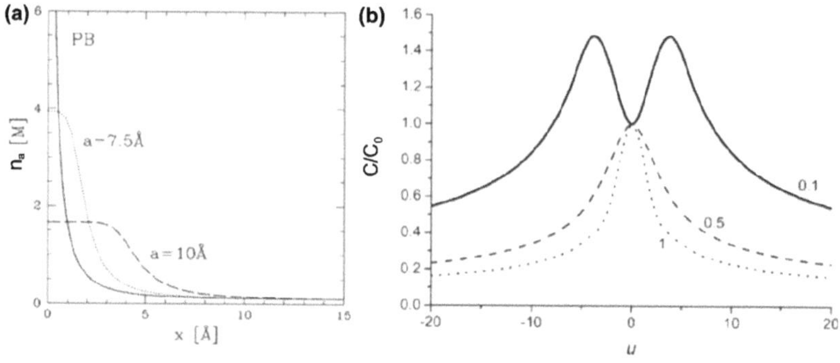

Figure 4.13 (a) Fermi-like modified Boltzmann distribution of ion density at the electrode electrolyte interface for different ion sizes. The solid curve illustrates the regular Poisson-Boltzmann description (denoted by PB) of the ion density. Reprinted with permission from [Borukhov et al. (1997)]. Copyright American Physical Society. (b) The EDL capacitance as a function of electrode potential which has taken into account the steric effect. Note that different values of γ was used. Reprinted with permission from [Kornyshev (2007)]. Copyright American Chemical Society.

Following the same derivation process from the previous subsection, the differential capacitance can be derived from the mPB equation by finding the stored charges on the electrode surface. This can be obtained by integrating Eq. 4.51 and applying the Gauss law. Therefore,

$$C_\mathrm{d} = C_0 \frac{\cosh\left(\frac{ze\varphi^m}{2k_\mathrm{B}T}\right)}{1+2\gamma\sinh^2\left(\frac{ze\varphi^m}{2k_\mathrm{B}T}\right)} \sqrt{\frac{2\gamma\sinh^2\left(\frac{ze\varphi^m}{2k_\mathrm{B}T}\right)}{\ln\left[1+2\gamma\sinh^2\left(\frac{ze\varphi^m}{2k_\mathrm{B}T}\right)\right]}}, \quad (4.52)$$

where $C_0 = \frac{\epsilon}{\lambda_\mathrm{D}}$, and λ_D is defined in Eq. 4.35. Figure 4.13b shows the calculated differential capacitance for different values of γ. It can be seen that when $\gamma = 0.1$, a "camel"-shaped behavior is obtained as that observed in experiments (Fig. 4.11). This suggests that the steric effect accounts for the decrease of C_d at high electrode potentials. It should also be noted that when γ increases, a maximum is observed at zero potential in the modified Poisson–Boltzmann theory, rather than a minimum in the GCS model. At high electric potentials, C_d is found to change with $\varphi^{-\frac{1}{2}}$. This anomalous behavior is important for the design and study of non-Faradaic electrochemical systems. In addition, if $\gamma = 0$, Eq. 4.52 becomes

$C_d = C_0 \cosh\left(\frac{ze\varphi^m}{2k_B T}\right)$, which is the same as Eq. 4.39. This is consistent with the previous conclusion that the modified Poisson–Boltzmann model returns to the GCS theory when $\gamma = 0$.

It is worth noting that, recent studies have shown that a short-range Coulombic correlation either among the ions [Bazant et al. (2011)] or between the ions and their image charges in the electrode [Skinner et al. (2010)] should be taken into account. This correlation is nonnegligible when the Bjerrum length[1] is larger than the ion diameter [Levin (2002)]. It leads to an overscreening effect for the solvent-free ionic liquids at small voltages, where the first ionic layer at the electrode–electrolyte interface consists of more counter-ions than the co-charges on the electrode. The layer next to this one in the electrolyte then sees a net charge of the opposite sign (to that on the electrode), thus again overscreen the first layer, and so forth, until the electroneutrality is reached in the bulk electrolyte. This structure for the EDL at the electrode–electrolyte interface could be used to explain the molecular layering model used for an experiment [Mezger et al. (2008)].

4.3.3 Ion Dynamics: The Vibration Model

The model introduced in the previous subsection is useful to investigate the electrochemical systems at equilibrium, as the minimum of the free energy was evaluated. This is useful for the study of the slow non-Faradaic processes where the transient evolution of EDL is not important. However, the recent development of electrochemical systems for various applications, such as energy storage and biosensing, requires a deeper understanding of the ion dynamics in EDL of non-equilibrium systems. This, for example, would help to design a better supercapacitor which can be made of both bulk and nano-porous materials. In this section, the dynamics of ions in EDL is studied to describe the steric effect.

The Nernst–Planck model along with the Poisson's equation are first introduced in the section to account for the steric effects on the electrode surface. The derived result describes well the

---

[1]The Bjerrum length $\lambda_B = \frac{(ze)^2}{4\pi\epsilon k_B T}$ defines a distance between two charges where their electrostatic energy is equal to the thermal energy.

decreasing differential capacitance at high electrode potential. This theory also predicts sharp peaks at potentials close to φ_{PZC}. These, however, are not observed in experiments. By performing a time-dependent study on ionic transport using the fluid mechanical equations, ions are found to vibrate near the electrode surface, in response to the applied electric field [Zhan et al. (2017)]. This vibration allows us to correctly predict the experimentally observed decreasing differential capacitance at high electrode potential. This new model elucidates the mechanism behind the ionic dynamics at solid–electrolyte interfaces, providing useful insight that may be applied to many electrochemical systems in energy storage, photoelectrochemical cells and bio-sensing. It is worth noting that, similar to the modified Poisson–Boltzmann theory, the ionic vibration model also predicts that the differential capacitance follows an inverse square root law at higher electrode potential. To simplify the study, the influence of the electrode surface condition, specific adsorption and the difference in ionic radii for anions and cations have been neglected.

4.3.3.1 Ion dynamics by the Nernst–Planck equation

As described in section 4.1.1, the Nernst–Planck equation can be employed here to study the ion dynamics for the steric effects in EDL. Due to the finite size of ions, there is a limit for the number of ions which can fit into the Stern layer (section 4.3.2). At high electrode potentials, the ion density increases drastically and hence reaches the limit. It is then straightforward to presume that the ions stacking on top of the original Stern layer to form a second layer, when the electrode potential is high enough [Attard (1995); Bazant et al. (2004); Bohinc et al. (2001); Kilic et al. (2007); Kornyshev (2007)], as illustrated in Fig. 4.14a. These layers (noted as sublayers in the rest of the chapter) comprise a dense region next to the electrode surface, which is called the **compact Stern layer** [Bazant et al. (2011)]. As a result, the location of OHP (x_2) moves away from the electrode surface with the increasing electrode potential (φ^m). Therefore, the effective Stern layer thickness, r_x, becomes greater than r_{H}, as shown in Fig. 4.14a. This in turn affects the differential capacitance.

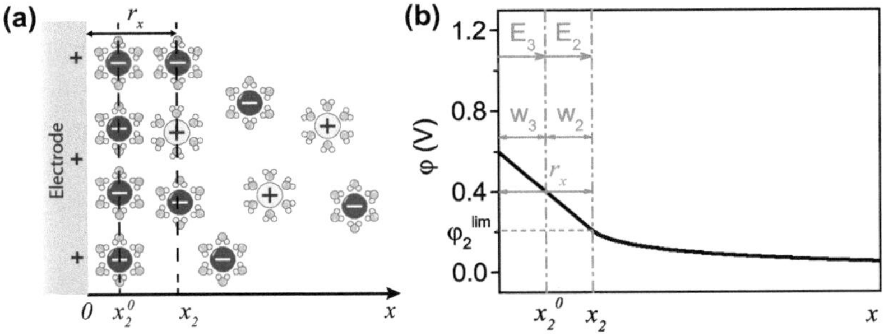

Figure 4.14 (a) The EDL structure including the compact Stern layer (the layering structure), where r_x, the distance from electrode surface to OHP, increases when the surface concentration reaches the limit. (b) Calculated potential distribution next to the electrode surface in electrolyte.

As discussed in the previous subsection, it is useful to find the relationship between φ and x (i.e., $-\epsilon\left(\frac{d\varphi}{dx}\right)_{x=0}$) in order to find the differential capacitance at various electrode potentials. The local electric field strength at position x in electrolyte ($-\frac{d\varphi}{dx}$) drives particles migrating in a certain direction. The Nernst–Planck model states that the particle flux in electrolyte must balance by both diffusion and migration (electric field-driven motion) of counter species (e.g., cations/anions in electrolyte) [Zhan et al. (2016); Zhan and Hu (2011); Zhan et al. (2011a)]. Therefore, the relationship between φ and x can be derived by studying the ion dynamics using the Nernst-Planck model with the Poisson equation (i.e., the Poisson–Nernst–Planck theory).

Figure 4.14b shows the calculated potential distribution in a 10 mM KF electrolyte for $\varphi^m = 0.6V$. As shown in Fig. 4.14a, the compact Stern layer (the whole region between x_2 and 0) in this case is composed of two sublayers (with indices 2 and 3), which covers the areas from x_2 to x_2^0, and from x_2^0 to 0, respectively. w_2 and w_3, E_2 and E_3, σ_2 and σ_3 are the thicknesses, electric field strengths, and conductivities for sublayers 2 and 3, respectively. In this section, $w_3 = r_H$ and $w_2 = r_x - r_H$, and the index 1 is reserved for the diffuse layer. The compact Stern layer can be viewed similar to a Stern layer in GCS theory, but with a different layer thickness r_x, instead of r_H.

As shown in appendix C, the complete form of Poisson–Nernst–Planck equations for compact Stern layer is complicated (Eq. C.3), therefore a few assumptions are made to simplify the calculation (see section C.1 for justification). When the effective Stern layer thickness (r_x) is smaller than the Debye length (λ_D), i.e., $r_x \ll c\lambda_D$ (where $c \sim \sqrt{\varphi^m}$), and $|\frac{ze\varphi}{k_B T}| \gg 1$, the Poisson–Nernst–Planck equations can be written as

$$\begin{pmatrix} \sigma_2 & -\sigma_3 \\ w_2 & w_3 \end{pmatrix} \cdot \begin{pmatrix} E_2 \\ E_3 \end{pmatrix} = \begin{pmatrix} 0 \\ \varphi^m - \varphi_2^{\text{lim}} \end{pmatrix}, \tag{4.53}$$

where φ_2^{lim} is the potential on new OHP. Note that this equation system is only valid at moderate electrode potentials. This is because when the electrode potential is too low ($|\frac{ze\varphi}{k_B T}| < 1$), there is no compact Stern layer. If the electrode potential is too high, the thickness of the compact Stern layer would exceed the Debye length, i.e., $r_x > c\lambda_D$.

For $\sigma_2 = \sigma_3$ the solution at high potential becomes $E_2 = E_3 = \frac{\varphi^m - \varphi_2^{\text{lim}}}{r_x} = -S^{\text{lim}}$, where S^{lim} is the slope of $\varphi(x)$ when $x < r_x$. The same result can be obtained numerically by Steepest Descent and Newton–Raphson methods. Note that the justification for the approximation $\sigma_2 = \sigma_3$ can be found in appendix D.

Since the ions distributed away from OHP ($x \in (x_2, \infty)$) still follow Maxwell–Boltzmann statistics, a relation similar to GCS theory can be written for the compact Stern layer as $\frac{\varphi^m - \varphi_2^{\text{lim}}}{r_x} = \sqrt{\frac{8n_0 k_B T}{\varepsilon}} \sinh\left(\frac{ze\varphi_2^{\text{lim}}}{2k_B T}\right)$. Hence the differential capacitance is expressed as

$$\frac{1}{C_d} = \frac{r_x}{\varepsilon} + \frac{1}{C_D^{\text{lim}}}, \tag{4.54}$$

where $C_D^{\text{lim}} = \sqrt{\frac{2\varepsilon z^2 e^2 n_0}{k_B T}} \cosh\left(\frac{ze\varphi_2^{\text{lim}}}{2k_B T}\right)$ is the capacitance of the diffuse layer. It can be proven that φ_2^{lim}, hence S^{lim} and C_D^{lim} are constant at high electrode potential because the total ionic areal density at OHP is always equal to n_{max}^{2D} when a compact Stern layer is formed (refer to appendix E for more details). Therefore,

$$r_x = \begin{cases} r_H & (n_{\text{OHP}}^{2D} < n_{\text{max}}^{2D}) \\ \dfrac{\varphi_2^{\text{lim}} - \varphi^m}{S^{\text{lim}}} & (n_{\text{OHP}}^{2D} = n_{\text{max}}^{2D}) \end{cases}, \tag{4.55}$$

where n_{OHP}^{2D} and n_{max}^{2D} are defined in section 4.3.2.

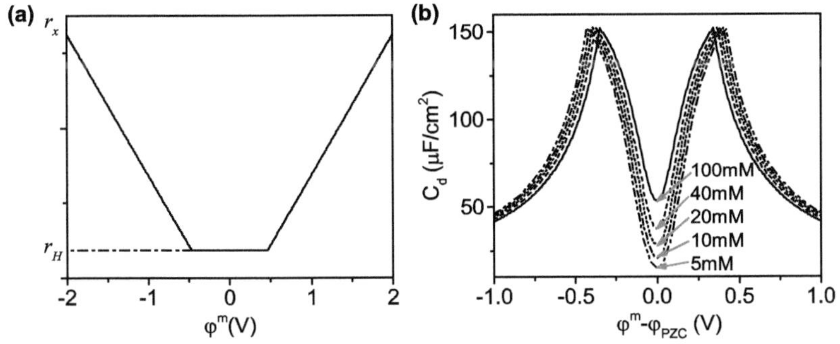

Figure 4.15 (a) r_x calculated by Eq. 4.55. (b) Differential capacitance of KF aqueous solution at different concentration, calculated by equation system 4.53.

Figure 4.15b shows the calculated differential capacitance as a function of electrode potential with respect to the PZC using Eqs. 4.54 and 4.55. The calculated C_d decreases at high potential as observed in the experiments [Islam et al. (2008); Lockett et al. (2008); Valette (1982)]. The evident sharp peaks in the maxima at low potential (around $\pm 0.4V$) have also been predicted in previous reports studying ionic dynamics in the compact Stern layer [Kilic et al. (2007)], which result from the turning points in r_x as shown in Fig. 4.15a. However these sharp peaks are not observed in experiments, indicating that the theory is inadequate at low potential. Note that a comparison of this Poisson–Nernst–Planck theory with the modified Poisson-Boltzmann equation, is presented in appendix B.

4.3.3.2 Fluid mechanics

It can be seen from the previous discussion that the Poisson–Nernst–Planck equation indeed could be used to account for the steric effect in EDL by including the compact Stern layer. However, it does not deliver a satisfactory result in comparison with the experiments, due to the occurrence of the sharp peaks. There are two reasons for this. (i) The approximations (e.g., $\frac{ze\varphi}{k_B T} \gg 1$ and $x \ll c\lambda_D$) used to obtain Eq. 4.53 are only valid at moderately high electrode potential. Despite this, these sharp turning points can still be observed after relaxing these approximations [Kilic et al. (2007)]. (ii) The Poisson–Nernst–Planck model (ambipolar diffusion) describes the ionic momentum transport at equilibrium where the time-dependent behavior is neglected [Zhan et al. (2016); Zhan and Hu (2011); Zhan et al. (2011a)]. This means that the model is suitable for high electrode potential, because the electrostatic force is large. However, the time dependence should be taken into account at low potential.

Similar to most diffusion phenomena investigations [Bird et al. (2007); Zhan and Hu (2011); Zhan et al. (2011a)], a mass transport (or continuous) equation is included in our study. Therefore, a system of equations describing the time-dependent behavior of the ions near the electrode surface can be written as

$$\frac{\partial n}{\partial t} + \frac{\partial}{\partial x}(nv) = 0, \tag{4.56a}$$

$$mn\frac{dv}{dt} = mn\left(\frac{\partial v}{\partial t} + v\frac{\partial v}{\partial x}\right) = -\frac{\partial P}{\partial x} + zenE, \tag{4.56b}$$

$$\frac{\partial E}{\partial x} = \frac{ze\,(n - n_0)}{\varepsilon}, \tag{4.56c}$$

where P and E are the pressure and the electric field in liquid, m, n and v are the mass, the density and the velocity of the ion, respectively. n_0 is the bulk ionic density at equilibrium. Equations 4.56a and 4.56b are the mass and momentum transport equations, respectively, and Eq. 4.56c is the Poisson's equation. Although equation system 4.56 is shown for a single component liquid, the equations and the results for a multicomponent solution (a binary system, for example) are very similar. The friction term is also neglected in Eq. 4.56b for simplicity.

It is convenient to decompose the ionic density into a bulk equilibrium term (n_0, independent on position x and time t) and a perturbation term on electrode surface (n_1, dependent on x and t), i.e., $n = n_0 + n_1$. Considering the pressure P as a constant in liquid along the direction x, can be reduced

$$\frac{\partial n_1}{\partial t} + n_0\frac{\partial v}{\partial x} = 0, \tag{4.57a}$$

$$m\frac{\partial v}{\partial t} = zeE, \tag{4.57b}$$

$$\frac{\partial E}{\partial x} = \frac{zen_1}{\varepsilon}. \tag{4.57c}$$

Rearranging all these equations gives

$$\frac{\partial^2 n_1}{\partial t^2} + n_1 \frac{n_0 (ze)^2}{\varepsilon m} = 0. \tag{4.58}$$

This is a typical harmonic oscillator equation. The perturbation ionic density can be written $n_1 = n_A \exp(-i\omega t)$, where n_A is the amplitude, and the oscillation frequency is $\omega = \pm\sqrt{\frac{n_0(ze)^2}{\varepsilon m}}$. Recall that the oscillation frequency for a harmonic oscillator is $\pm\sqrt{\frac{\eta}{m}}$, we then have $\eta = \frac{n(ze)^2}{\varepsilon}$ as a force constant. This indicates the ionic density near the electrode surface varies harmonically, while an individual ion vibrates, in response to the electric field. It should also be noted that although the oscillation frequency is only dependent on n_0 and m, the amplitude (and hence the vibration energy) is expected to change with the electric field (potential). This then serves as a basis for our **vibration model** for ions.

4.3.3.3 Ion vibration in electrical double layer

In essence, the classical radius of an elementary particle (such as atom or ion) is the result of the short-range interatomic forces. The elementary particles could approach to each other until they "feel" the Lennard–Jones potential [Kittel (2004)]. Similarly, in the presence of the electric field, the Coulomb potential determines the closest distance where the ions can approach each other (or the electrode surface). This is manifested by the previously discussed vibration effect. Hence, r_x as the average closest distance that the ions can approach the electrode surface in EDL is redefined in this section.

Vibration models have often been adopted to study intermolecular forces [Kittel (2004); Marcus (1956)]. Our model proposed previously describes Coulomb ionic interactions, providing a better way to calculate the "compact Stern layer" thickness. In the presence of electric field, ions will be driven towards the electrode. When they are close enough to the surface, all their potential energy will be converted into kinetic energy and they will bounce back and forth due to Coulombic and short-range intermolecular forces, as shown

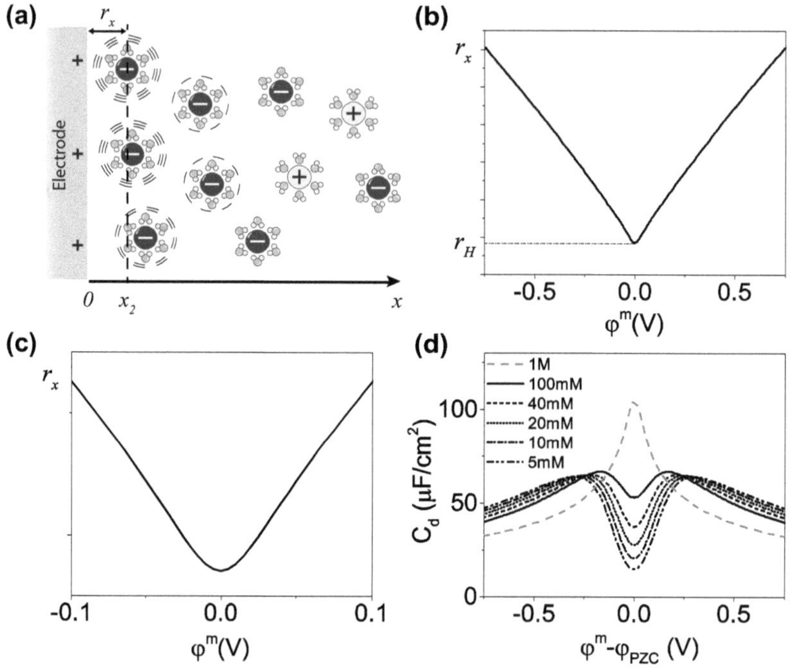

Figure 4.16 (a) The EDL structure including the vibration model, where r_x, the distance from electrode surface to OHP, increases when the surface concentration reaches the limit. The dashed circles surrounding the solvated ions denote the ionic vibration. (b) r_x calculated by Eq. 4.62. (c) Calculated r_x in smaller scale indicates the smooth transition across PZC. (d) Calculated differential capacitance of KF aqueous solution at different concentration by the vibration model.

in Fig. 4.16a. A real-life analogy is that of dropping a basketball on solid ground. Hence, the energy due to harmonic oscillation can be expressed as $\frac{1}{2}\eta(\Delta r)^2$, where Δr represents the oscillation amplitude.

Before the discussion of the energy conservation at electrode–electrolyte interfaces, it is useful to review thermodynamics and fluid mechanics. The particles in liquid at thermal equilibrium have the thermal energy $\frac{3}{2}k_B T$ due to the random motion of the particles at a given temperature (T). An isothermal system with a well-defined temperature, however, does not necessarily reach thermodynamic equilibrium. This is because other state variables (other than temperature) may still change with time. Therefore, thermodynamic processes are usually discussed in electrochemistry [Bard and Faulkner (2000); Hassibi et al. (2004)]. In electrode–electrolyte systems, the thermal energy ($\frac{3}{2}k_{\mathrm{B}}T$) mainly contributes to the thermal noise of the measured signals [Hassibi et al. (2004); Nyquist (1928); Zeitler and Fromherz (2013)]. This indicates that the energy of ions contributed by other dynamical processes (such as diffusion, electrical migration, or harmonic oscillation) should be taken into account. Fluid mechanics indicates that ion transport in liquid can also occur subject to the energy provided by other factors [Kundu and Cohen (2010)], where the energy conservation can be obtained by taking the 2nd order moment of the Boltzmann transport equation (section A.3). This is written as

$$n\frac{\mathrm{d}}{\mathrm{d}t}\left(\frac{3}{2}k_{\mathrm{B}}T+\frac{1}{2}mu^2\right)=-\nabla\cdot(\vec{Q}+\overset{\leftrightarrow}{P}\cdot\vec{u})+n\vec{F}\cdot\vec{u}, \qquad (4.59)$$

where $\frac{3}{2}k_{\mathrm{B}}T$ is the thermal energy given by the equipartition theorem, $\frac{1}{2}mu^2$ is the kinetic energy, $\vec{u}$ is the mean velocity, $\vec{Q}$ is the heat flux, $\overset{\leftrightarrow}{P}$ is a general stress tensor (which can be considered as pressure for simplicity), $\vec{F}$ is the external force. In this case, $\vec{F}$ is the electrostatic force, $\vec{F}=q\vec{E}$. For an electrolyte where the temperature is constant, we have $\frac{\mathrm{d}}{\mathrm{d}t}\left(\frac{3}{2}k_{\mathrm{B}}T\right)=0$. Therefore, Eq. 4.59 becomes

$$\Delta\left(\frac{1}{2}n\eta\Delta r^2\right)=\Delta\mathcal{E}+n\vec{F}\cdot\vec{u}\Delta t, \qquad (4.60)$$

where $\mathcal{E}$ represents the energy of ions in electrolyte in other form. Note that $\mathcal{E}$ not only contains the first right-hand-side term in Eq. 4.59, but may also include, for example, the ion-ion correlation energy [Bazant et al. (2011)] and re-solvation energy if the electrode is sub-nanoporous [Kondrat and Kornyshev (2011)]. $\frac{1}{2}mu^2$ is replaced by an energy term due to harmonic oscillation ($\frac{1}{2}\eta\Delta r^2$). This is because the time-dependent ionic transport in this case manifests itself as harmonic oscillation, as previously discussed.

Therefore, the general form of the energy conservation equation can be written as

$$\frac{1}{2}n_t\eta(r_x-r_1)^2+n_t\mathcal{E}_1=\frac{1}{2}n_t\eta(r_{\mathrm{H}}-r_0)^2+n_t\mathcal{E}_0+|n_{\mathrm{net}}ze\Delta\varphi|, \qquad (4.61)$$

where $n_{\text{net}} = n_c - n_a$, η is the force constant, and $\Delta\varphi$ is the potential drop. The left-hand side in the equation denotes the total ionic energy when the electrode potential $\varphi^m \neq \varphi_{\text{PZC}}$, while the first two terms in the right-hand side are the total ionic energy at PZC, and the third term is the electrostatic energy of the ions. $\mathcal{E}_1$ and $\mathcal{E}_0$ are the ionic energies in other forms with and without electric field, respectively. r_x and r_{H} are the 'compact Stern layer' thicknesses, r_1 and r_0 are the closest distances an ion approaches to the electrode surface, when the electrode is biased at φ^m and φ_{PZC}, respectively. Hence $r_x - r_1$ and $r_{\text{H}} - r_0$ represent the corresponding amplitudes of the vibration. If the electric field does not affect the ionic energy in other forms, i.e. $\mathcal{E}_1 = \mathcal{E}_0$, the solution of r_x is obtained

$$r_x = \sqrt{(r_{\text{H}} - r_0)^2 + \left|\frac{ze}{\eta}\tanh\left(\frac{ze\varphi_x}{k_{\text{B}}T}\right)\Delta\varphi\right|} + r_1, \tag{4.62}$$

where $\varphi_x = \varphi(r_x)$, $r_1 = r_0$ since $r_x(\Delta\varphi = 0) = r_{\text{H}}$, and a relation $\frac{n_{\text{net}}}{n_t} = \frac{n_c - n_a}{n_c + n_a} = \tanh\left(\frac{ze\varphi_x}{k_{\text{B}}T}\right)$ has been used. It can be seen from Fig. 4.16b and that there are no sharp turning points in the calculated r_x using the vibration model. Figure 4.16c indicates that the transition of r_x across PZC ($\varphi^m = \varphi_{\text{PZC}}$) is smooth. It is also worth noting that r_x curve is nearly linear at moderately high potential, as described by the Poisson–Nernst–Planck model (Fig. 4.15a). Since $\Delta\varphi$ increases with the electrode potential, a relation $\Delta\varphi \sim \alpha\varphi^m$ is used to model the differential capacitance in this section.

Figure 4.16d shows the calculated capacitance for a KF electrolyte at different bulk concentrations (ranging from 5 mM to 100 mM). It can be seen that the shoulder of the "camel"-shaped curve gets closer to the minimum of PZC compared to the curve depicted in Fig. 4.15b, or in other words there is an evident decrease in the capacitance value. This is because the OHP gradually moves away from the electrode surface as the ionic vibration becomes stronger at higher potential. The shoulder also gets wider and lower as the bulk electrolyte concentration decreases. The reason is that the diffuse layer capacitance becomes more dominant at lower bulk concentration. All these phenomena have been observed in past experiments as shown in Fig. 4.11d and are modeled well by the vibration model.

At high electrode potential, $\tanh\left(\frac{ze\varphi_x}{k_B T}\right) \approx 1$; hence $C_d \sim r_x^{-1} \sim (\varphi^m)^{-\frac{1}{2}}$. This relation was also confirmed by an mPB theory (section 4.3.2). The gray dashed line in Fig. 4.16d is the differential capacitance for an electrode in a 1 M KF solution. The curve has an interesting bell shape indicating that the diffuse layer capacitance has only a minor effect on the total differential capacitance. Note that this is also consistent with the mPB theory for high surface ion concentration even in dynamic processes (i.e., γ has a large value in Fig. 4.13b).

It has been experimentally shown that the double layer structure has a temperature dependence [Lockett et al. (2008); Mezger et al. (2008)]. Although a layering structure has been used to model the result, it cannot explain the temperature dependence. Similar to the vibration model for Einstein solids [Kittel (2004)], the presented ionic vibration model suggests that r_x should be dependent on temperature when the ionic energies in other forms are taken into account. This requires further developments in this theory.

Based on the deep understanding of EDL capacitance obtained in this section, we are able to design experiments which include solid electrode in electrolyte, potentially for the application of biosensing and/or energy storage. Electrodes with large capacitance are generally useful for biological application such as neuron stimulation or capacitive biosensing. The conventional GCS theory suggests that a higher capacitance value could be achieved by increasing the electrode potential. However, the new ion vibration model [Zhan et al. (2017)] concludes that it might be impossible to increase the capacitance simply by applying a higher potential. In addition to the development of electrodes with large surface area, we may have to carefully select the electrical potential, ion concentration, ion size, ionic interactions, and oscillation frequency to optimize the capacitance.

4.4 Summary

This chapter briefly introduces the theories used to study the electron transfer and ion dynamic processes at electrode–electrolyte interfaces. Several different theories ranging from the classical Nernst–Planck equation to the most recent ion vibration model are discussed.

- The (Poisson–) Nernst–Planck equation (Eq. 4.5) is useful to discuss the ion transport at steady state in electrolyte, and it is simplified from the fluid mechanics.
- The electrochemical potential in electrolyte (Eq. 4.11) is equivalent to the Fermi level in solid crystals.
- An electric potential in electrolyte will be screened by counter-ions, and hence it decays rapidly with the distance in the solution. At a certain distance (screening length), the electric potential decreases to a value which is only e^{-1} of the original one. When the ions follow the Maxwell–Boltzmann distribution, the screening length is called the Debye length (Eq. 4.13).
- The activation energy of a redox reaction can be calculated by the Marcus theory where the Gibbs free energy of the ions are calculated using an ion vibration model (Eq. 4.20).
- The electron transfer rate can be calculated using the Arrhenius equation (Eq. 4.21) with the activation energy derived by the Marcus theory.
- The Gerischer model proposes that the oxidized and reduced species are analogous to the conduction and valance bands (the unoccupied and occupied states) in semiconductors (Fig. 4.5).
- The electric current for Faradaic processes (redox reactions) can also be calculated using the Gerischer model where the properties of the electrode are taken into account (Eqs. 4.28–4.31).
- Non-Faradaic processes are conventionally described by the Gouy–Chapman–Stern theory where the electric double layer (consists of the Stern layer and diffuse layer) is defined (Fig. 4.9a).
- The electric potential profile in the electric double layer is described by Eqs. 4.40 and 4.42, and the differential capacitance is expressed by Eq. 4.44 (Fig. 4.10b). Note that the equivalent circuit of the electric double layer is the capacitors contributed by the Stern layer and diffuse layer in series.
- The steric effect of the electric double layer at high surface ion concentrations can be modeled by the modified Poisson–Boltzmann equation Eq. 4.51.

- The modified Poisson–Boltzmann equation is obtained by minimizing the free energy at equilibrium, and therefore is not suitable for the investigation of the transient behavior of ion dynamics.
- The ion vibration model (Eqs. 4.58 and 4.58) derived from the time-dependent fluid mechanics equations not only accounts for the steric effect, but also describes the ion dynamics at electrode–electrolyte interfaces for non-equilibrium systems (Fig. 4.16).

Chapter 5

Graphene–Electrolyte Systems

At graphene–electrolyte interfaces, the molecules/ions in electrolyte interact intimately with electrons, defects, and carbon atoms in graphene on the microscopic level. This interaction would induce doping effects or/and introduce charged impurities, and hence it changes the electronic properties of graphene, including the density of states, the Fermi level, and/or carrier mobility. If the electrolyte is simply regarded as a dielectric medium on the macroscopic level, a dielectric screening effect which enhances the carrier mobility in graphene must be considered. These changes in the electronic properties of graphene are important to the fundamental electronic processes at graphene–electrolyte interfaces, such as charge transfer discussed in section 4.2. Studies on these topics not only offer opportunities to engineer redox reactions, but also provide a guideline to improve graphene electronic devices. In the first section of this chapter, we discuss briefly the changes in electronic properties of graphene due to several molecule-graphene interactions, which are categorized differently, by reviewing typical examples from recent research.

For certain applications involving graphene–electrolyte systems, such as biosensing, neuron stimulation/recording, and energy storage, the information of the interfacial capacitance is as important

Graphene–Electrolyte Interfaces: Electronic Properties and Applications
Hualin Zhan

ISBN 978-981-4774-94-9 (Hardcover), 978-1-003-04487-1 (eBook)
www.jennystanford.com

as the charge transfer. In the remaining sections of this chapter, we use the knowledge from the previous chapters to discuss the band alignment processes in solid-electrolyte systems which involve the interfacial capacitance and the density of states. An ion–electron heterostructure is constructed in analogy to the electronic heterostructures in section 2.3, because the electrolyte could be investigated as a type of special semiconductor (i.e., the Gerischer model in section 4.1.3). This model is commonly used in cutting-edge research, and it is very useful for both fundamental studies and application development.

5.1 Physisorption and Chemisorption

Depending on the types of bonding formation between the particles (molecules, atoms, or ions) and the carbon atoms in graphene, the types of adsorption of particles on graphene can be categorized into physisorption and chemisorption. If the particles stay only on the surface of graphene due to van der Waals interactions, they are being physisorbed (i.e., **physisorption**). In contrast, if a covalent bonding or an ionic bonding is formed between the particles and graphene, this adsorption process is then defined as **chemisorption**. In essence, all these absorption processes are due to the electric forces (one of the four elementary interactions in the universe). However, the **van der Waals interaction** is weak as it is *mainly contributed by the electric attraction between the induced dipole moments of the particles* (where the repulsion force comes from the Pauli exclusion principle). **Covalent bonding** or **ionic bonding** is much stronger, because *either the particles share electrons or there is a direct Coulomb force between the charged ions* [Kittel (2004)].

5.1.1 First-Principle Calculation and Doping

When in contact with chemical species in electrolyte, graphene is chemically doped due to the physisorption and chemisorption processes. The electronic properties of graphene, such as the local density of states (LDOS), the Fermi level ($\mathcal{E}_F$), and mobility, change

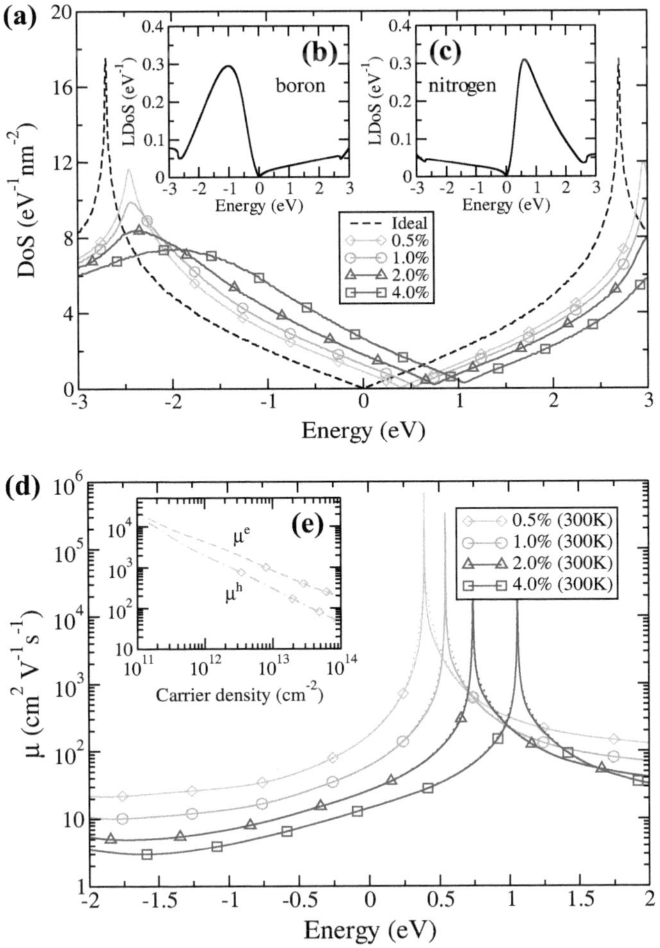

Figure 5.1 (a) DOS of a boron-doped (as heteroatom) graphene for different doping concentrations. The local DOS around a (b) boron and (c) nitrogen dopant. (d) Charge carrier mobility (μ) in graphene as a function of the energy for different doping concentrations. (e) The dependence of electron (μ^e) and hole (μ^h) mobilities on the charge carrier density for the doping concentration of 5%. Reprinted with permission from Lherbier et al. (2008) Copyright American Physical Society.

because of the electric interaction between the chemical species and the carbon atoms in graphene. For physisorption, the induced dipole moments play an important role due to the nature of the van der Waals interaction. When covalent or ionic bondings are formed in chemisorption, the stronger electric interaction indicates that the charge transfer process should be considered. As the energy of elementary particles (e.g., electrons) satisfies the Schrödinger equation (Eq. 2.12), the aforementioned interaction between the chemical species and graphene can be described by solving the Schrödinger equation. It should be noted that the exact expression of the general solution of Eq. 2.12 for an arbitrary potential is usually difficult to obtain. Therefore, a common technique used to describe the chemical adsorption processes is to find the numerical solutions of the Schrödinger equation using modern numerical methods, for example, the density functional theory. This method is often referred as the **first-principle (or ab initio) calculation**.

Figure 5.1 demonstrates the ab initio calculations of the doping effect (i.e., the change of the electronic properties) for the chemisorption of boron and nitrogen atoms in graphene. The external atoms/molecules used to induce the doping effect are defined as **dopants**. Note that in this type of chemisorption, the carbon atoms in graphene are substituted by the boron (or nitrogen) atoms (i.e., the dopant atoms are heteroatoms). Figure 5.1a shows the shift in the Dirac point and the change in the overall density of states (DOS) for a boron-doped graphene. It can be seen that the Dirac point shifts to the positive direction while $\mathcal{E}_F$ is at 0. This indicates that the graphene sheet is hole-rich (refer to γ in Fig. 3.5a). This is straightforward to understand, as boron (a group III element) acts like an acceptor of electrons (i.e., donor of holes) in the carbon (a group IV element) lattice. It, then, can be inferred that the nitrogen (a group V element) dopant would induce an n-type doping (electron-rich) of graphene. In addition, the variation of the LDOS close to the boron and nitrogen dopants in graphene is shown in Figs. 5.1b and 5.1c, respectively. This indicates that the LDOS varies drastically from the overall DOS for a doped graphene. It is worth noting that the mobilities of electrons and holes have different values for the chemically doped graphene, as shown in Figs. 5.1d–e. This information is useful for the charge carrier transport in the presence of a magnetic field, as discussed in detail in Chapter 8.

When a molecule is covalently bonded to graphene, the electronic properties change drastically as the atoms in the molecules

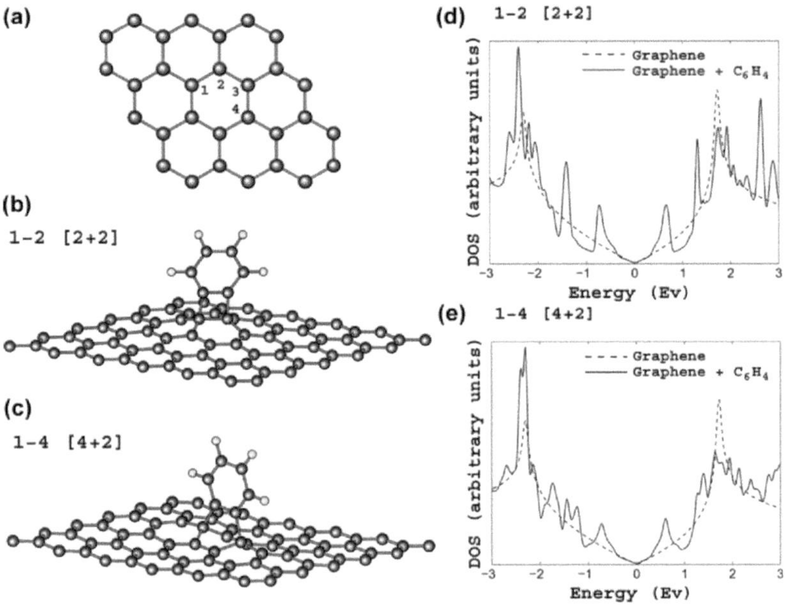

Figure 5.2 (a) The graphene lattice showing the numbered positions for the attachment of benzyne molecules to a hexagonal carbon ring. (b, c) Benzyne molecules are attached to graphene sheets in different positions to form the covalent bonding, where the [2+2] and [4+2] cycloaddition reactions are considered, respectively. (d, e) The calculated density of states for the functionalized graphene sketched in (b) and (c), respectively. Reprinted with permission from Hammouri et al. (2015) Copyright American Chemical Society.

interact strongly with those in graphene. Because of their complex structures, molecules can attach to graphene in various spatial configurations. Figures 5.2a–c show different functionalization configurations of a benzyne molecule bonded on graphene. The first-principle calculation confirms that the density of states of a chemisorbed (covalently bounded) graphene changes significantly from that of a pristine one, as indicated in Figs. 5.2d–e. This information is particularly useful to study the electronic structure of graphene when its surface is chemically modified (functional groups are chemisorbed on graphene), i.e., **surface functionalization**. Graphene oxide introduced in section 6.1.2 is an example of functionalized graphene. As can be inferred from Fig. 5.2, the calculated band structure of graphene oxide is significantly different from that of pristine graphene (Fig. 3.4), and it is largely dependent on the composition, position, and orientation of the functional groups (molecules) on graphene [Mkhoyan et al. (2009)].

The conventional density-functional theory methods using either local density approximation (LDA) [Perdew and Zunger (1981)] or the generalized gradient approximation (GGA) [Perdew et al. (1996, 1992)] fail to describe the van der Waals interaction, as they do not hold for the discussion of electronic interactions across the regions where the electron density is very low. Therefore, new functionals need to be added into the simulation for physisorption. Popular techniques to account for the van der Waals interaction include adopting the London dispersion formula [Ortmann et al. (2005)] or calculating the local and non-local correlation energy of electrons [Dion et al. (2004)]. Recent studies combining the advantages of these methods [Tkatchenko and Scheffler (2009)] are used to calculate the interaction between some molecules, such as the deoxyribonucleic acid (DNA) bases, and graphene.

Figures 5.3a–b show the cross-sectional view for guanine and adenine absorbed on graphene, where the contour maps of the electron density are illustrated. It is shown that the electron density between the DNA bases and graphene is low, confirming the weak interaction for physisorption. The distance between the absorbed molecules and graphene is determined by finding the minimum energy of the system, as indicated in Fig. 5.3c. The calculated electronic structures of graphene with guanine (G), adenine (A), thymine (T), and cytosine (C) absorbed are demonstrated in Fig. 5.3d. Note that the lowest unoccupied molecular orbital (LUMO) and the highest occupied molecular orbital (HOMO) are also illustrated by the upper and bottom blue thick lines, respectively. It can be seen that the band structure of graphene near the "Dirac" point is almost unaffected by the physisorbed molecules. This is because the van der Waals interaction is weak, as previously discussed.

The ab initio calculations for physisorption and chemisorption of chemical species on graphene indicate that the interaction between the adsorbed atoms/chemicals and graphene is highly dependent not only on the composition and structure of the species, but also on its orientation on graphene and the form of

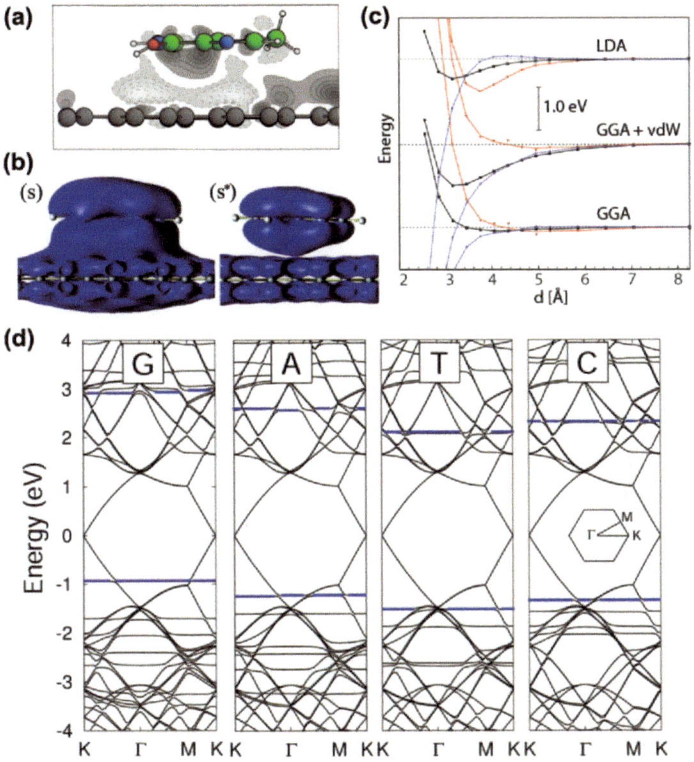

Figure 5.3 (a) and (b) The contour maps of electron density for guanine (a) and adenine (b) adsorbed on graphene. The bonding (s) and anti-bonding (s*) combinations are shown in (b). The outmost isodensity line in (a) is 4×10^{-4} Å^{-3}, while the surface in (b) is 10^{-6} Å^{-3}. (c) The total (black), exchange-correlation (blue), and kinetic (red) energy are plotted as a function of the distance between adenine and graphene, calculated using different approximations. (d) Electronic band structure of absorbed DNA bases on graphene. The LUMO and HOMO are indicated by the thick blue lines. (a) and (d) are reprinted with permission from Lee et al. (2013b) Copyright American Chemical Society. (b) and (c) are reprinted with permission from Ortmann et al. (2005) Copyright American Physical Society.

bondings. It is then useful to perform the first-principle calculation for graphene–electrolyte system to investigate the fundamental interaction between the chemical species in solution and graphene.

However, since the number of ions/molecules (including those for solvent) which should be considered in the calculation is very large and the ions/molecules keep moving at the interface, the first-principle calculation for such a system requires a tremendous amount of resources. This is not practical for most applications. Therefore, a common method taking into account the electron (Fermi and vacuum) energy to draw the band diagram will be introduced in section 5.2.

5.1.2 Dielectric Screening

Graphene, as an atomically thin carbon material, is sensitive to its environment [Adam et al. (2007); Castro Neto et al. (2009); Schedin et al. (2007)]. The ions and defects in the surroundings affect the electronic properties of graphene, such as the charge carrier density and the mobility, through a number of different ways. As discussed previously, the ions and defects can dope the graphene by changing the local density of states [Freitas et al. (2011); Kong et al. (2014); Leenaerts et al. (2008, 2009); Lherbier et al. (2008); Pi et al. (2010); Singh et al. (2013); Wehling et al. (2008)], and hence modulate the carrier density. They are also able to change the carrier mobility by introducing charged impurity (Coulombic) scattering (discussed in section 3.2.3.2) [Adam et al. (2007); Newaz et al. (2012)] and dopants [Lherbier et al. (2008)]. The changes in these quantities can then be captured by Hall effect measurements (Chapter 8). There are a few aspects which should be taken into account when performing the electrical measurements of a semiconductor (graphene) in solution, such as the effects of dielectric screening and electrical double layer structure.

In addition to the influence of charged impurities, Coulombic scattering in a semiconductor is also affected by the dielectric constant of the environment through the dielectric screening effect [Chen et al. (2009a,b); Farmer et al. (2009); Jang et al. (2008); Jena and Konar (2007); Ponomarenko et al. (2009); Siegel et al. (2013); Xia et al. (2010b)]. The high dielectric constant in the environment changes the Thomas–Fermi screening length of charges in the material. Hence, the scattering rate is reduced, and the mobility is enhanced.

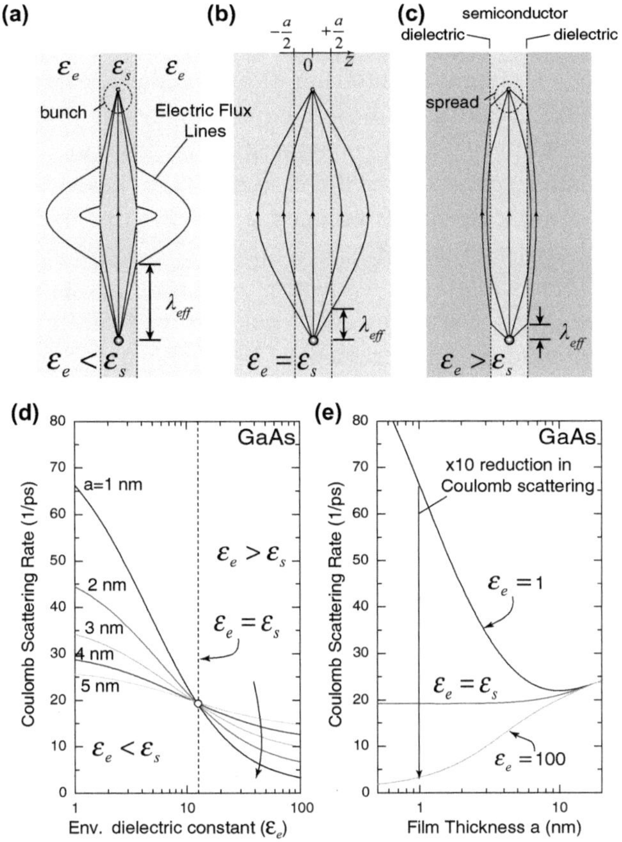

Figure 5.4 Electric field lines emerging from a charged impurity (the bottom sphere) towards a mobile electron (the top dot) in a semiconductor which is in a dielectric environment with (a) $\varepsilon_e < \varepsilon_s$, (b) $\varepsilon_e = \varepsilon_s$, and (c) $\varepsilon_e > \varepsilon_s$. ε_e and ε_s are the dielectric constants of the dielectric and the semiconductor, respectively. a is the thicknesses, of the semiconductor, and λ_{eff} is a rough indication for the effective Thomas–Fermi screening length. Coulombic scattering rate as a function of (d) the dielectric constant for a semiconductor (GaAs) with different thickness, and (e) the semiconductor (GaAs) thickness with various dielectric constants. Reprinted with permission from Jena and Konar (2007). Copyright American Physical Society.

As shown in Figs. 5.4a–c, the electric field lines close to the electron (the top dot) spread out as the dielectric constant of the environment (ϵ_e) increases. This means the Coulombic force that the electron experiences gets weaker, and hence the Thomas–Fermi screening length (roughly indicated by λ_{eff}) is shorter. This in turn reduces the Coulombic scattering rate, as indicated in Figs. 5.4d–e. Therefore, in addition to the charged impurities present in the environment, the dielectric constant of the environment also affects the Coulombic scattering. In the case of a high dielectric constant, the Coulombic scattering is reduced, and hence the mobility of the semiconductor is enhanced.

The influence of this dielectric screening on the electrical properties of graphene, however, could be twofold. In addition to the aforementioned enhancement of the mobility, it should also be noted that, due to the reduced screening of the impurity potential by conduction electrons, the carrier-density-independent conductivity limited by the short-range scattering could be decreased by, for example, 38% [Jang et al. (2008)].

5.2 Band Alignment Involving Electrolytes

The microscopic study of the interaction between graphene and the chemical species in electrolyte is important, as it provides the essential information for physical processes. When designing electrochemical systems, it is useful to relate these microscopic interactions to the quantities measurable, such as the interfacial capacitance or Faradaic current. As discussed in sections 2.3.2 and 4.2, band alignment at the interface of two materials is a useful technique for designing the systems. The band bending in the band diagram could reflect some of these interactions, as it results from the redistribution of charges at the interface.

5.2.1 Metal–Electrolyte Systems

In general, carrier densities in semiconductors, electrolytes, and metals are of the order of 10^{16} cm^{-3}, 10^{20} cm^{-3}, and 10^{22} cm^{-3}, respectively [Myamlin and Yurii (1963)]. This indicates that the screening lengths in these phases (W) are related like so, $W_s \gg W_e \gg W_m$, due to Eq. 4.13, where the subscripts s, e, and m represent semiconductor, electrolyte, and metal, respectively. Therefore, a rough estimation using $C \sim \frac{\epsilon}{W}$ gives the relation of the capacitances in these phases, resulting in $C_s \ll C_e \ll C_m$. In a heterostructure, the number of charges in each phase is the same at the interface, and hence the corresponding potential drops have this relation: $\varphi_s \gg \varphi_e \gg \varphi_m$. These relationships are particularly useful for the band alignment in heterostructures involving an electrolyte.

An electrolyte is similar to a moderately doped semiconductor to an extent, as the charge carriers in both follow a Boltzmann distribution. However, there is no bandgap in an electrolyte. In this regard, an electrolyte is similar to a metal, which also does not have bandgap, except that the charge carriers are described by a different distribution function. The band diagram alignment for a metal–electrolyte heterostructure is illustrated in Fig. 5.5 using the four-step technique introduced in section 2.3.2 and the previous conclusions. Figures 5.5a and 5.5c show the electron energy levels before contact for metals with smaller and larger work functions than electrolyte, respectively. Figures 5.5b and 5.5d depict the potential distribution (red) across the interface. It can be seen that the potential difference across the interface distributes mostly in the electrolyte, because of the distinct difference between the charge carrier density. This potential drop is generally referred to as the electrochemical potential of a metal in an electrolyte in electrochemistry theory [Bard and Faulkner (2000); Memming (2015)], and the corresponding capacitance in the electrolyte at the metal–electrolyte interface is the double layer capacitance (section 4.3).

5.2.2 Semiconductor–Electrolyte Systems and Photoelectrochemistry

Band diagram alignment for a (p-type) semiconductor–electrolyte system is shown as an example in Figs. 5.5e–h. As demonstrated in Figs. 5.5e–f, the work function of electrons in a semiconductor is greater than in an electrolyte, while in Figs. 5.5g–h it is lower. The different values of the work function determine the energy profile formed at the interface, i.e. the formation of either Schottky barrier or Ohmic contact similar to those in Fig. 2.20. Figures 5.5e and 5.5g show the energy levels before contact, while Figs. 5.5f

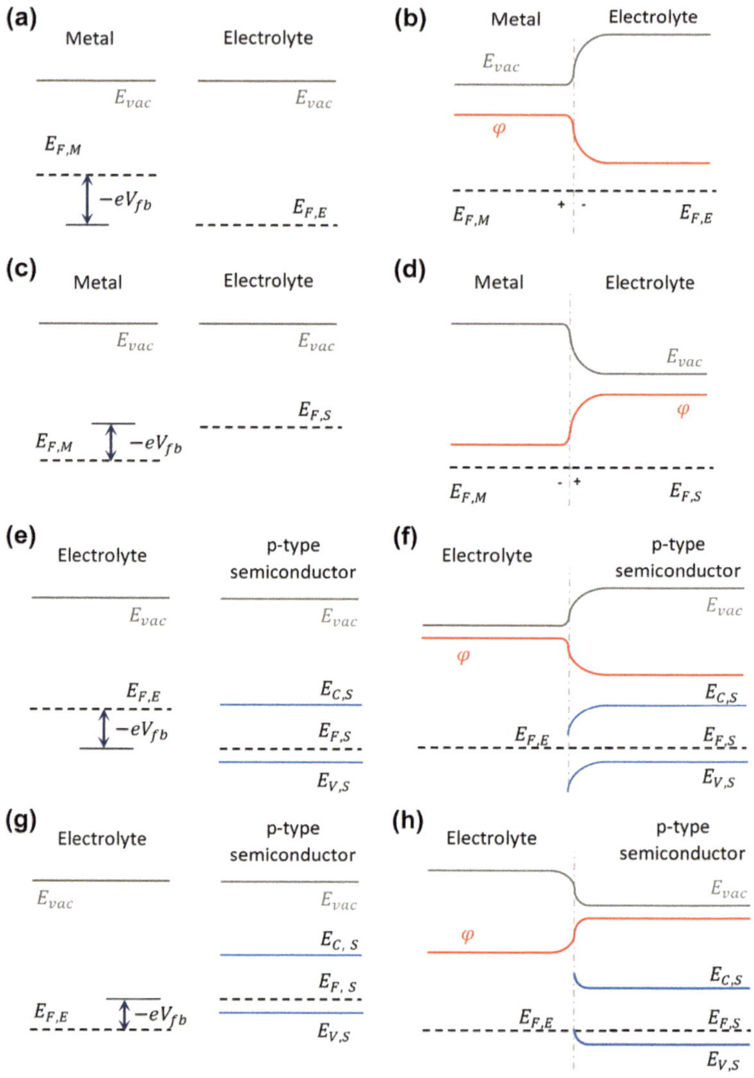

Figure 5.5 Band diagrams of (a–d), a metal–electrolyte heterostructure, and (e–h), a (p-type) semiconductor–electrolyte heterostructure.

and 5.5h illustrate the band diagrams and the potential distribution (red) across the interface after contact. As the charge carrier concentration is generally lower in semiconductors, when compared to electrolytes, the value of W_{dep} is larger in semiconductors. Therefore, the potential drop in such a system mostly occurs in semiconductors (when a depletion layer forms, as shown in Fig. 5.5f). Hence, the capacitance at the interface is usually dominated by the depletion layer capacitance (or defined as semiconductor capacitance in some literature) in semiconductors. A Mott–Schottky relation can then be derived by solving the Poisson equation, which is useful for estimating the doping level for semiconductors.

However, it should be noted that this potential distribution profile may change dramatically when an accumulation or inversion layer is formed at the interface of the semiconductor (Fig. 5.5h), as the degenerate charge carrier density in these layers generally has a much larger value [Davies (1998); Szkopek (2013)]. In this case, the double layer capacitance dominates the total capacitance, as if the semiconductor were a metal. This information is particularly useful for electrochemical measurements using semiconductor electrodes, such as silicon, silicon carbide, diamond, and other such materials.

Now let us turn to capacitance measurements of a metal–electrolyte and a graphite–electrolyte system as an example for previous discussions. Figure 5.6a shows the density of states of a metal and graphite. Since the density of states of graphite is low, it is considered as a semiconductor in graphite–electrolyte systems. This gives a lower charge density at the surface of graphite than that of electrolyte or metal, as shown in Fig. 5.6b. It can also be seen that the charges in graphite are distributed in a large space, due to the previously discussed long screening length. This indicates that the semiconductor capacitance of graphite (C_{SC}) is lower than the capacitance of the electrical double layer [Gerischer (1990)]. Note that the electrical double layer (EDL) consists of a Helmholtz layer (where the capacitance is C_H) and a diffuse layer (where its capacitance is neglected in the cited work), and hence C_{SC} is dominant in a circuit where C_{SC} is in parallel with C_H, as indicated by the measured total differential capacitance (C_{exp}) in Fig. 5.6c.

Photoelectrochemistry is an important application of semiconductor–electrolyte systems. When bandgaps are present in semiconductors, a depletion layer can form at the semiconductor–electrolyte interface for the materials such as silicon, silicon

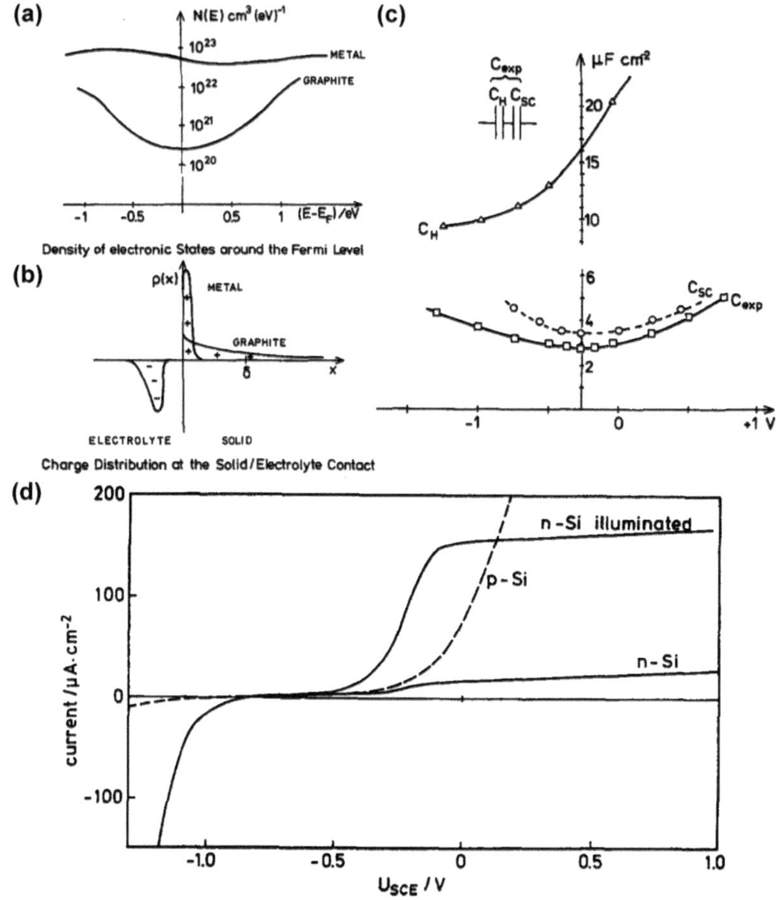

Figure 5.6 (a) Comparison of the density of states for a metal and graphite near the Fermi level. (b) Illustration of charge distribution at metal–/graphite–electrolyte interfaces. (c) Measured differential capacitance (C_{exp}) at a (stress-annealed) graphite–electrolyte interface, where C_H is the capacitance of the Stern (Helmholtz) layer in electrical double layer, and C_{SC} represents the semiconductor capacitance of the graphite. (d) Photocurrent at an n-type silicon–electrolyte interface (solid lines), in comparison with the current at the p-type silicon–electrolyte system. Reprinted with permission from Gerischer (1990) Copyright Elsevier.

carbide, diamond, among others. When light is shone on the depletion layer formed at the semiconductor–electrolyte interface, the electrical current in the system rises rapidly, as shown in Fig. 5.6d. This process not only modulates the Faradaic current, but also changes the interfacial capacitance. This photocurent opens up a new field of photoelectrochemistry [Bard et al. (2002); Gellings and Bouwmeester (1997); Gerischer (1990); Memming (2015)], and it is useful for various applications including photovoltaic cells, neuron stimulation, photocatalysis, and so on.

It is also worth noting that, in a semiconductor–electrolyte system, the potential drop within the semiconductor may be similar to that within the electrolyte at the interface. This is because the carrier density in the semiconductor is much lower than that in metals. In the case of a heavily doped semiconductor or/and low concentrated analyte, the carrier densities in semiconductor and solution might be comparable. As a result, both the potential drops within the semiconductor and the electrolyte should be taken into account. Therefore, a relation between the gate voltage and all the potential drops can be written as

$$V_g = V_{fb} + \varphi_{EDL} + \varphi_s, \tag{5.1}$$

where V_g, V_{fb}, and φ_s are the potential drop across the interface, the flat-band voltage, and the potential in semiconductor due to the band bending, as illustrated in Figs. 5.5e–h. These quantities are similar to those defined in section 2.3.2.1. φ_{EDL} represents the potential within the electrolyte (electrical double layer, EDL) at the interface. Note that there is no oxide layer in this system. Therefore, φ_{ox} from Eq. 2.36 is removed.

5.3 Graphene–Electrolyte Systems

Graphene, as an atomically thin semimetal, has a unique gapless band structure which makes the energy diagram of a graphene–electrolyte system distinct from those mentioned earlier.

First of all, it is difficult to draw the energy band edge as a function of position in the direction perpendicular to the graphene sheet, because there is no band edge in the band structure and

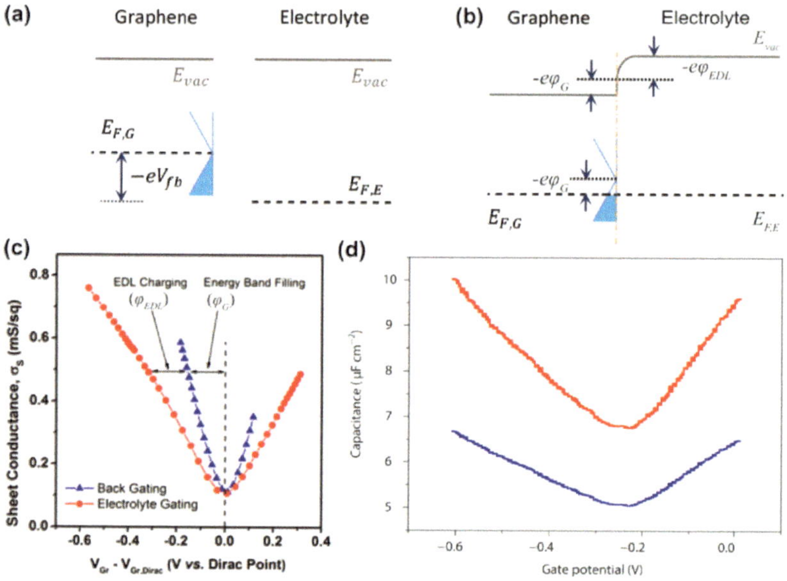

Figure 5.7 Band diagrams (a) before and (b) after graphene contact electrolyte. (c) Measurement of the potential drops in graphene (φ_G) and an electrolyte (φ_{EDL}) using a dual-gate transistor structure. Reprinted with permission from Kim and Frisbie (2014). Copyright American Chemical Society. (d) Graphene quantum capacitance (red) in comparison with the total capacitance (blue), measured using an electrochemical cell where a momolayer graphene was adopted as the working electrode. Reprinted with permission from Xia et al. (2009). Copyright Nature Publishing Group.

2D graphene sheets are atomically thin. Therefore, unlike the conventional band diagrams shown previously, the band diagram of a graphene–electrolyte system may be plotted using the energy with respect to the density of states, as shown in Fig. 5.7a.

Secondly, due to the absence of a bandgap, no depletion layer will be formed in graphene. As a result, there is no depletion layer capacitance in graphene, in contrast to regular semiconductors where it will be formed. The double layer capacitance might be dominant in the electrical measurement of such a system if graphene behaves as a metal and has a large charge carrier density.

Lastly, the carrier density in graphene ($\sim 10^{12}$ cm^{-2}) is lower than that in a thin slab of electrolyte ($\sim 10^{14}$ cm^{-2}) on some occasions. As the number of charges stored in the EDL at the graphene–electrolyte interface cannot exceed that of the counter-charges in graphene, the quantum capacitance of graphene should be taken into account. Research suggests that the Fermi degeneracy energy gives a non-negligible potential drop in graphene (φ_G) as compared with that found in an electrolyte (φ_{EDL}), due to the small value of quantum capacitance [Ji et al. (2014); Xia et al. (2009)], as illustrated in Fig. 5.7b. Note that the quantum capacitance also gives a partial penetration effect of the electric field [Luryi (1988)], which should be considered in sandwich-like structures where graphene is in the middle [Georgiou et al. (2013)].

An experimental measurement using a dual-gate transistor structure confirms that φ_G and φ_{EDL} are comparable in some cases, as demonstrated in Fig. 5.7c. This indicates that both φ_G and φ_{EDL} need to be considered when calculating the potential drop across the graphene–electrolyte interface (V_g), which is proportional to the difference between the Fermi level in graphene and the electrochemical potential in the electrolyte. Since the potential due to band bending (φ_s) in graphene is usually not considered because of the monoatomic thickness, an equation similar to Eq. 5.1 can be written for a graphene–electrolyte system:

$$V_g = V_{fb} + \varphi_{EDL} + \varphi_G, \tag{5.2}$$

where V_{fb} is the flat-band voltage shown in Fig. 5.7a. $-eV_{fb}$ is the energy difference between the Fermi level in graphene and the electrochemical potential in electrolyte before graphene is in contact with the electrolyte, when the vacuum levels are aligned.

This discussion is useful to study the processes at graphene–electrolyte interfaces, such as estimating the doping level, deriving the interfacial capacitance, and calculating the Faradaic current. The example shown in Fig. 5.7b indicates that the graphene sheet is p-doped when it is in contact with the electrolyte. The quantum and EDL capacitance values can then be derived by finding the derivatives of the charge carrier concentration with respect to φ_G and φ_{EDL}, respectively. Figure 5.7d shows the measured quantum (red) and total (blue) capacitance of a graphene–electrolyte system using a three-electrode electrochemical cell setup. Note that the quantum capacitance is in the same order of magnitude with the total capacitance, although the quantum capacitance is greater. This suggests that the quantum capacitance can affect the total capacitance, and it should be taken into account in some graphene–electrolyte systems.

As both the EDL and the quantum capacitances vary with the voltage, it is important to choose an experimental technique for capacitance measurement, where the voltage does not change too much during one measurement (for a specific potential). Therefore, a range of different experimental techniques for graphene and electrochemistry are briefly reviewed in the following chapters.

5.4 Summary

The physical processes which can affect the electronic properties at graphene–electrolyte interfaces are briefly summarized in this chapter. The research methods such as ab initio simulation and band diagram analysis are also introduced.

- Chemical species can attach to graphene by forming chemical bonds (chemisorption) or through the van der Waals interaction (physisorption).
- Depending on whether the process is chemisorption or physisorption, the electronic properties of graphene, including the charge carrier concentration and/or mobility, could be affected differently through doping effects or induced charged impurities.
- These interaction processes can be studied by solving the Schrödinger equation using first-principle simulations.
- If graphene is placed in a medium with high dielectric constant, the charge carrier mobility could be greatly enhanced. This is because the Coulomb scattering rate is reduced due to the dielectric screening effect.
- The charge carrier concentration in electrolyte is generally higher than that in semiconductors and lower than that in metals. This makes the band diagram for systems involving electrolyte very different (Fig. 5.5).
- The depletion layer formed at semiconductor–electrolyte interfaces might give a smaller capacitance (the semiconductor

capacitance) than the EDL capacitance, which suggests that the semiconductor capacitance must be considered in semiconductor–electrolyte systems.

- The light-induced electron–hole pairs generated in the depletion layer lead to a photocurrent, which opens up the field of photoelectrochemistry (Fig. 5.6).
- The atomically thin nature of graphene suggests that there is no depletion layer capacitance. However, the quantum capacitance of graphene should be considered in some cases due to the low density of states.

Chapter 6

Experimental Methods for Graphene

This chapter presents a survey on the popular experimental techniques of graphene. We start from reviewing the most common fabrication methods of monolayer graphene, and then move onto the characterization techniques. By obtaining an overview on these topics, we are able to select the fabrication technique which produces graphene samples of desired quality for various purposes. In addition to verifying the quality, some of the characterization techniques, such as Raman spectroscopy, can also be used to investigate the electronic and photonic properties of graphene. Note that in many of the fundamental and application research of graphene, additional investigation of electron transport is required. Therefore, in the final sections of this chapter, the popular electrical measurement techniques such as the four-probe method and the Hall effect measurement are introduced.

6.1 Growth Techniques

It is useful to review the existing growth techniques of graphene before we use it in an application, as this will enable us to select the samples with the appropriate quality for different purposes. For

Graphene–Electrolyte Interfaces: Electronic Properties and Applications
Hualin Zhan

ISBN 978-981-4774-94-9 (Hardcover), 978-1-003-04487-1 (eBook)
www.jennystanford.com

example, the graphene that we need to use for biosensing should have higher quality than that for energy storage applications (e.g., supercapacitors). This is because the selectivity and the sensitivity for the detection of molecules are of primary concern; and graphene with higher structural and electronic quality is more sensitive to the environment than those with defects.

6.1.1 Mechanical Cleavage

The first experimental evidence of the atomically thin carbon layer (graphene) was given by K. Novoselov and A. Geim using the **mechanical cleavage** (MC) method [Novoselov et al. (2004)]. This famous Scotch tape method simply exfoliate the monolayer of carbon from graphite (usually highly oriented pyrolytic graphite, HOPG). It produces graphene sheets with the highest structural and electronic quality up to present. Figure 6.1a shows a high-resolution microscopic study of a graphene sheet obtained by MC technique [Park et al. (2010)]. Region B is concluded containing a monolayer graphene due to the stronger intensity of the $\{1\bar{1}00\}$ spot. The high

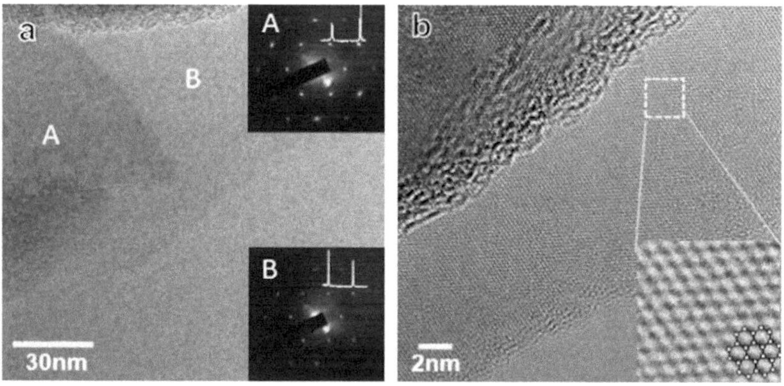

Figure 6.1 (a) A low-magnification transmission electron microscopy (TEM) image showing a multilayer (A) and a monolayer (B) region on a graphene obtained by MC. The insets show the diffraction pattern and the intensity ratio between the $\{1\bar{1}00\}$ and $\{11\bar{2}0\}$ spots for region A and B. (b) A high-resolution TEM image of a monolayer graphene and its edge. The honeycomb lattice is clearly shown. Reprinted with permission from [Park et al. (2010)] Copyright Elsevier.

magnification of the TEM image demonstrated in Fig. 6.1b clearly indicates the hexagonal lattice structure.

In MC experiments, an HOPG was first treated with oxygen plasma to create a micron-size (typically 5 μm) mesa. This mesa was pressed into a photoresist, and then baked. In the following step, the Scotch tape was used to peel the carbon flakes off the sample. Finally, the flakes were placed in acetone, and the Si/SiO_2 substrate was used to capture them. A 1 mm large MC created graphene sheet was reported in 2009, where the electronic mobility was 15000 $cm^2V^{-1}s^{-1}$ [Geim (2009)].

6.1.2 Liquid Phase Exfoliation

Even with the help of oxygen plasma treatment, MC method is still time-consuming, which is not suitable for large-scale production. In addition, the size of graphene obtained by MC is not large. The liquid phase exfoliation (LPE) method provides a facile alternative to produce large multi-layered graphene sheets or membrane by creating reduced graphene oxide (rGO) dispersion, due to its excellent colloidal stability and the liquid-phase processability.

Figure 6.2a shows the fabrication process of rGO colloids. An HOPG is first oxidized in liquid to obtained the graphite oxide using the Hummers method [Hummers and Offeman (1958)], where the inter-layer spacing of graphite oxide is larger than that of graphite. This material is then sonicated to create graphene oxide (GO) colloids, which is followed by a reduction step to obtain the graphene dispersion. Multilayered graphene membrane can then be assembled from the dispersion via wet processing methods including casting, filtration, and spinning. Depending on the reduction methods, the obtained graphene sheets may or may not form irreversible agglomerates [Li et al. (2008); McAllister et al. (2007)], as demonstrated in Fig. 6.2b. The agglomerate-free dispersion is critical to synthesize multilayered graphene membranes with better structural uniformity. It should be noted that the graphene sheets produced by LPE may still have functional groups due to the incomplete reduction process. A transmission electron microscopy image indicated in Fig. 6.2c confirms that atoms other than carbon are also present in rGO. This indicates that LPE

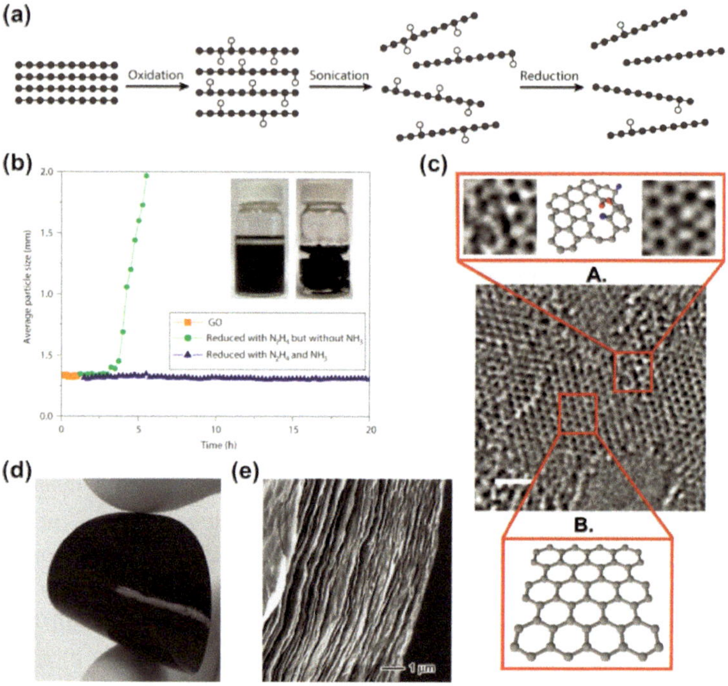

Figure 6.2 (a) Illustration of the liquid phase exfoliation process of graphene. An HOPG is first oxidized in liquid to obtained graphite oxide with a larger inter-layer spacing. The material is then sonicated to create graphene oxide (GO) colloids. This is then reduced chemically to obtain the graphene dispersion. (b) Average particle size as a function of time for GO and reduced graphene oxide (rGO, reduced with and without ammonia). Inset: the photos of rGO which is reduced with (left) and without (right) ammonia. Reprinted with permission from [Li et al. (2008)] Copyright Nature Publishing Group. (c) A TEM image of an rGO sheet (scale bar: 1 nm). Oxygen (red balls) and nitrogen (blue balls) are present in the proposed atomic structure of region A, while perfect carbon atomic lattice is observed in region B. Reprinted with permission from [Erickson et al. (2010)] Copyright Wiley-VCH. (d) An optical photo of a multi-layered rGO membrane. Reprinted with permission from [Yang et al. (2011)] Copyright Wiley-VCH. (e) A scanning electron microscopy image of an rGO showing the layered structure. Reprinted with permission from [Chen et al. (2008a)] Copyright Wiley-VCH.

is useful when the quality is not of great concern and the large-scale production is necessary. Note that various choices of reduction method in LPE, on the other hand, offer a unique advantage of multilayered graphene membrane, where the surface chemistry could be controlled. Another advantage of LPE is that a flexible multi-layered rGO membrane can be readily obtained, as shown in Figs. 6.2d and e. This structure is useful for applications in energy storage and water treatment where, for example, the construction of nano-sized spaces for ions and molecules is important.

It is worth noting that although the conventional LPE technique requires the oxidation of graphite to increase the interlayer distance between the graphene layers, recent advances in this technique indicate that it is possible to apply the LPE on graphite directly using microwaves and molecularly engineered ionic liquids [Matsumoto et al. (2015)].

6.1.3 Chemical Vapor Deposition and Plasma-Enhanced Chemical Vapor Deposition

Chemical vapor deposition (CVD) is one of the most popular methods used for the mass production of high-quality graphene. This technique produces graphene sheets on metal substrates using carbon in the gas form, where the substrate can also be used as the catalyst. Nickel was used in this method in 2008 [Yu et al. (2008)], while copper (Cu) was then reported as a substrate for large-area CVD growth of graphene in 2009 [Li et al. (2009)]. The mobility for the graphene grown on Cu was around 4050 $cm^2V^{-1}s^{-1}$. With the help of surface oxygen on the substrate, this number was increased to the range between 15000 to 30000 $cm^2V^{-1}s^{-1}$ at room temperature when graphene is then transfered to hexagonal boron nitride (h-BN) substrates [Hao et al. (2013)]. This indicates that quality of CVD grown graphene is approaching to that made by MC. In addition, the roll-to-roll technique of the CVD growth of graphene on Cu foils provides a promising future for the industry-level production of high quality graphene [Bae et al. (2010)].

The CVD process can be catalyst-free with the assistance of plasma [Wang et al. (2004)], which can be generated by a number of different methods, such as microwave, electron cyclotron resonance, surface plasma, and so on. Hence this method is called plasma enhanced chemical vapor deposition (PECVD). This indicates that graphene could be grown on various substrates, since the metal catalyst is no longer required. As the carbon source comes from the gas (usually methane/hydrogen mixture), graphene can grow on substrates in complicated shapes by PECVD. It is reported recently that graphene can conformally grow on ordered nanoporous structures by PECVD [Zhan et al. (2016)]. This method gives a three-dimensional (3D) graphene network with a high quality and ordered nanoporous structures, where chemical methods mainly produce disordered or quasi-ordered porous rGO membrane with relatively lower structural and electronic quality. The obtained 3D graphene is promising for the applications not only in energy storage, neuron stimulation, and biosensing, but also in photonics due to the ordered porous structure.

6.1.4 Molecular Beam Epitaxy and Thermal Annealing of SiC

There are also other techniques for fabricating graphene, for example, molecular beam epitaxy (MBE) and thermal annealing of silicon carbide (SiC). MBE allows to produce graphene layer by layer from carbon-rich gas (e.g., ethylene) at high temperature (between 800 and 1200°C) [Zhan et al. (2011b)]. The electronic quality, however, is low, as the measured mobility is around 123 $cm^2V^{-1}s^{-1}$.

Thermal annealing of SiC is a method commonly used in scientific research to obtain high quality graphene. In this method, SiC is heated up to a certain temperature (typically between 1000°C and 1500°C) in vacuum (the temperature could be higher in inert gas, e.g. 1700°C in argon). This thermal decomposition process on the (0001) surface of 6H-SiC sublimate silicon (Si) leaves the carbon atoms to reconstruct into graphene layers. The mobility of the obtained graphene at room temperature ranges between 1100 and 9500 $cm^2V^{-1}s^{-1}$ [Berger et al. (2004); Zhang et al. (2017)], which can be further increased to 11000 $cm^2V^{-1}s^{-1}$ at the temperature of 0.3 K [Pallecchi et al. (2014)]. This indicates that the electronic quality of graphene obtained by this method is approaching that produced by CVD. It was reported that the 100 GHz transistors made of the graphene grown on SiC exceeds the state-of-the-art silicon transistors for high frequency applications [Lin et al. (2010)]. The wafer-scale production of graphene on SiC provides an alternative for future industry development [Emtsev et al. (2009)]. It should be noted that the buffer layer formed between the obtained graphene layer and the SiC substrate hinders the electrical performance. Therefore, a current interest of the research in this field is to find a facile technique to remove this buffer layer either during or after the thermal annealing process.

6.1.5 Comparison of Growth Techniques

Figure 6.3 compares the aforementioned fabrication techniques of graphene, where the electrical mobility, and structural and electronic quality are shown. This map enables us to select the appropriate fabrication method to suit our needs for different applications. For example, MC is ideal for investigation of novel electronic transport phenomenon in graphene; LPE can be used for energy storage systems due to the low-cost, facile fabrication process of nanoporous structures; CVD and SiC are suitable for development of next generation electronics and biosensors which require both high quality and the potential for mass production. In addition to the monolayer or few-layered graphene production, multi-layered graphene membranes with tunable interlayer spacing (nanochannels) can be fabricated by LPE [Yang et al. (2013)]. This indicates that a range of fabrication techniques of graphene are available for research beyond electronics, such as nanofluidics.

6.2 General Methods for Characterization

The quality of graphene varies drastically with the selected method and the parameter used for growth process. Therefore, it is important to characterize the obtained graphene, ensuring that the quality of the sample is sufficient for the experiments. The structure of graphene are usually examined by transmission

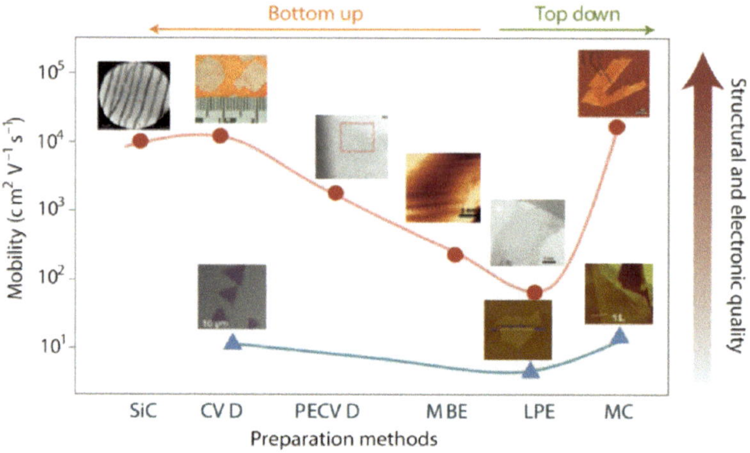

Figure 6.3 Carrier mobility of graphene (red curve) and MoS_2 (blue curve) as a function of preparation method, where CVD, PECVD, MBE, LPE, and MC stand for chemical vapor deposition, plasma-enhanced chemical vapor deposition, molecular beam epitaxy, liquid phase exfoliation, and micromechanical cleavage. Note that SiC represent a graphene growth technique by annealing silicon carbide substrate. Types of images from left to right for graphene: Low-energy electron microscopy image for a graphene on SiC; optical image for a CVD grown graphene on Cu; high-resolution transmission electron microscopy (HRTEM) image for a graphene by PECVD; scanning transmission microscopy (SEM) image for a graphene by MBE; HRTEM image for graphene films by LPE; and an atomic force microscopy (AFM) image for a graphene by MC. Types of images from left to right for MoS_2: optical photograph of a MoS_2 by CVD; AFM image for a MoS_2 by LPE; AFM image of a MoS_2 by MC. Reprinted with permission from [Fiori et al. (2014)] Copyright Nature Publishing Group.

electron microscopy (TEM) and atomic force microscopy (AFM), in addition to optical microscopy. A facile method which has been widely used to qualitatively characterize the overall quality of graphene is Raman spectroscopy, which includes investigating the atomic orbital hybridization (i.e., sp^2/sp^3 ratio), number of layers, defects, and grain size. In addition, the atomic orbital hybridization of carbon can also be determined by the electron energy loss spectroscopy (EELS) near the carbon K-edge. If the X-ray photons are used in this study, in stead of electron, the technique is referred as the near edge X-ray absorption fine spectroscopy (NEXAFS). The angle-resolved photoemission spectroscopy (ARPES) is useful for providing the information of the band structure of graphene.

6.2.1 Transmission Electron Microscopy and Atomic Force Microscopy

In **transmission electron microscopy**, an energetic electron beam is accelerated by the high electrical voltage provided by an electron gun. This is then irradiated on a thin slice of a sample, and the transmitted electron beam is focused and collected by a sensor to obtain the microscopic image. This process is similar to that of an optical microscope, where the wavelengths in both cases determine the resolution. As the wavelength (λ) of the electron reduces with the increasing energy, the measurable smallest feature (d) is determined by

$$d \approx \frac{\lambda}{2A_N} \approx \frac{0.6}{A_N\sqrt{V}}\ \text{nm}, \tag{6.1}$$

where A_N is the numerical aperture of the system and V is the voltage for acceleration. In theory, d could reach the sub-atomic level for a voltage of 100 kV. The real resolution, however, will be affected by the physical limitations of the TEM system. It should be noted that $d \approx 0.1$ nm is already achievable.

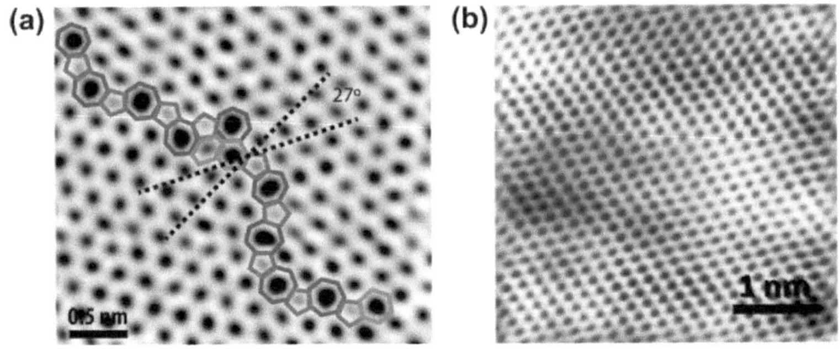

Figure 6.4 (a) A TEM image of a graphene grown on Cu by CVD, where the honeycomb lattice and the grain boundary are shown. Reprinted with permission from [Huang et al. (2011)] Copyright Nature Publishing Group. (b) An AFM image of a graphene grown by MBE. Reprinted with permission from [Lin et al. (2012)] Copyright American Institute of Physics.

Note that the use of TEM is sometimes limited by its complicated process of sample preparation. The working principle for **atomic force microscopy**, on the other hand, is much simpler. A mechanical tip is presented in the very close proximity of a sample, and the atomic force between the tip and sample is measured (typically by laser). The morphology of the sample can then be obtained, and the resolution is relatively lower than that of TEM. Figure 6.4 shows the TEM and AFM images of graphene, where the honeycomb lattice and the grain boundary can be clearly observed.

6.2.2 Raman Spectroscopy

Raman spectroscopy is a nondestructive method to investigate the structural and electronic properties of a material. It detects the Raman effects where an incident photon is scattered inelastically, accompanied by the production or annihilation of a phonon or magnon [Kittel (2004)]. During this process, the created or annihilated phonon deforms the primitive cell. This in turn changes the electric dipole moment due to the strain-dependence of the electronic polarizability. Hence, the photon is emitted with the modified frequency.

Owing to the advantages of this technique, Raman spectroscopy is one of the most popular tools for the study of carbon materials, such as diamond, amorphous carbon, carbon nanotube, graphite, graphene, and so on. In addition, Raman spectroscopy is particularly useful to graphene because the gapless electronic band structure allows the detection of incident radiation resonant at all wavelengths. This enables the technique to collect information about both the atomic structure and the electronic properties of graphene. Therefore, we are able to determine the number of atomic layers and the quality of graphene (including defects, disorder, etc.) through Raman measurements.

In general, D, G and 2D peaks in Raman spectra, as labeled in Fig. 6.5a, are most commonly used to study the quality of graphene, including determining the number of layers, defects, grain sizes, doping, and so on. The breathing modes of six-atom rings are responsible for the D peak which is activated in the presence of defect. Note that the lattice symmetry breaking in graphene will

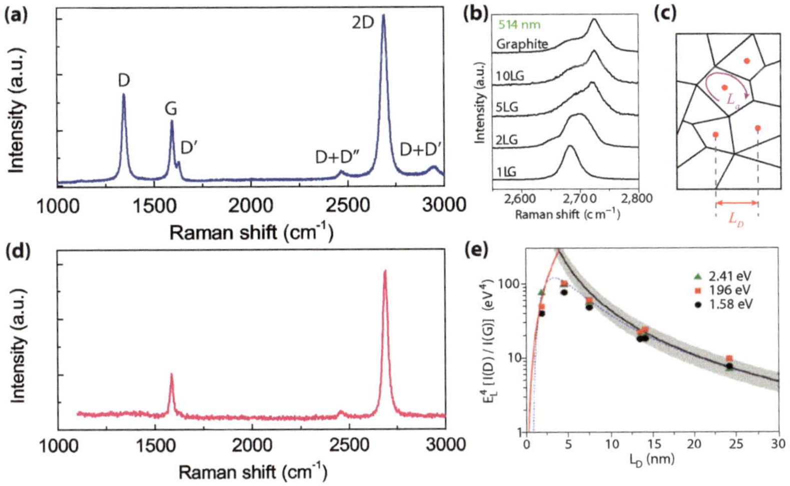

Figure 6.5 Raman spectra for (a) an ordered 3D graphene network fabricated by PECVD [Zhan et al. (2016)] and (d) a CVD graphene transferred to SiO_2. (b) 2D peak as a function of the number of graphene layers. (c) Illustration of a polycrystalline 2D material, where the fragments are the grains, and the red dots represent the defects. L_a and L_D are the grain size and the inter-defects distance, respectively. (e) The relation between $I(D)/I(G)$ and the distance between defects in graphene (L_D). (b) and (e) are reprinted with permission from [Ferrari and Basko (2013)] Copyright Nature Publishing Group.

give rise to the D and D′ peaks, where D′ peak corresponds to the double resonance happening as intravalley process. It is also worth noting that the D, D′ and D″ peaks are activated by one-phonon defect-assisted processes, while the D + D′ peak is produced by two-phonon defect-assisted processes. The G peak comes from the high frequency E_{2g} phonon at the Γ point. It increases at high doping level and can be modified near the graphene edge. Therefore the intensity ratio between D and G peaks are usually used to assess the defects and grain size in graphene and other carbon products [Ferrari and Basko (2013); Ferrari and Robertson (2000)]. 2D peak is the overtone of D peak, and it is usually sensitive to the number of layers due to its relation with the electronic band structure.

Figure 6.5a shows a Raman spectrum of an ordered 3D graphene network obtained by PECVD [Zhan et al. (2016)]. The Raman spectroscopic studies were conducted on a Renishaw in Via Raman Microscope with a laser wavelength of 532 nm and a laser spot size of 10 μm. The D, G and 2D peaks in G-AAO are located at 1348 cm^{-1}, 1595 cm^{-1}, and 2691 cm^{-1}, respectively. The full-width-at-half-maximum of the 2D peaks (FWHM(2D)) is generally used as an indicator of the number of graphene layers in sp^2 carbon, as shown in Fig. 6.5b. As the number of layers increases, FWHM(2D) broadens. It can be seen from Fig. 6.5a that the FWHM(2D) of the sample is 41.9 cm^{-1}. This is slightly larger than that of a CVD monolayer graphene on SiO_2 (32.3 cm^{-1}, as depicted in Fig. 6.5d) consisting of a single 2D component but lower than of bilayer graphene (50 cm^{-1}), which contains four 2D peak components [Ferrari et al. (2006)]. This suggests that the 3D graphene network is composed predominantly of monolayer graphene and the slight broadening of the 2D graphene peak could be explained by the defective nature of 3D graphene and a large number of grains in the focal spot size. The much greater intensity of 2D peak with respect to the G peak is also consistent with a dominant contribution from single layer graphene or rotated multilayer (non Bernal stacked) graphene flakes [Ferrari and Robertson (2000); Li et al. (2014a); Xiao et al. (2012)].

The D and D′ (1629 cm^{-1}) peaks suggest the existence of defects in the sample [Ferrari and Basko (2013)]. The intensity ratio between D and G peaks ($I(D)/I(G)$) is 1.5, which indicates disorder and the presence of sp^3 type defects [Ferrari and Basko (2013); Ferrari and Robertson (2000); Matthews et al. (1999); Tuinstra and Koenig (1970)], most probably originating from grain boundaries in the graphene layers. As mentioned earlier, the intensity ratio between D and G peaks is a common method used to analyze disorder in graphitic samples [Knight and White (1988); Matthews et al. (1999); Tuinstra and Koenig (1970)]. For the 532 nm laser excitation, the average graphene grain size is estimated using an empirical model as

$$L_a = \frac{C}{I(D)/I(G)} = \frac{4.96}{I(D)/I(G)} \approx 3.3 \text{ nm}, \tag{6.2}$$

where $C \approx 0.033\lambda_L - 12.6$ nm is a constant linearly proportional to the laser wavelength (λ_L). Figure 6.5e shows $E_L^4[I(D)/I(G)]$ as a function of L_D using a similar relation, where E_L is the laser excitation energy [Ferrari and Basko (2013)]. $L_D \sim \sqrt{L_a}$ is the inter-defects distance, as illustrated in Fig. 6.5c.

6.2.3 Electron Energy Loss Spectroscopy

It is worth noting that during the TEM characterization of a sample, we are able to identify the chemical element of the material. This is done by analyzing the energy loss of the electron beam focused on a specific position on the sample, and the method is defined as the electron energy loss spectroscopy (EELS). It can be applied to investigate the sp^2/sp^3 ratio of a graphene sample.

Figure 6.6a shows carbon K-edge EELS spectra for a 3D graphene network obtained by PECVD [Zhan et al. (2016)], while 6.6b is for a glassy carbon, using a focused electron beam ($\approx$0.5 nm). The EELS characterizations were conducted on a JEOL 2100F. The electron energy applied in EELS studies was 80 keV, and the collection angle was 16 mRad. The peak evident at the energy-loss of 285 eV in Fig. 6.6a represents the π^* resonance in carbon K-edge, which corresponds to a transition of electrons to unoccupied carbon π^*

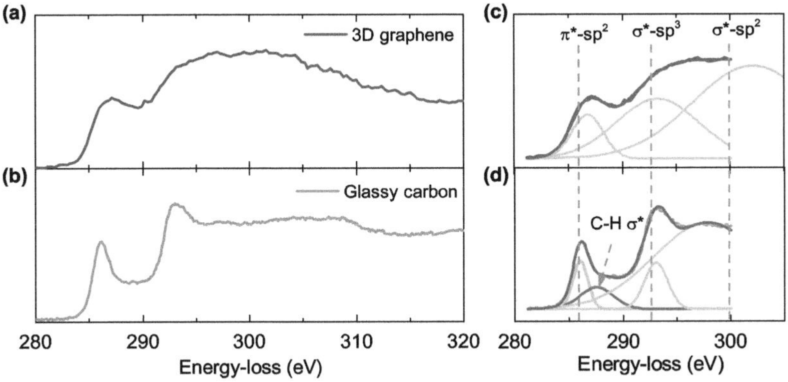

Figure 6.6 EELS spectra of (a) a 3D graphene network obtained by PECVD [Zhan et al. (2016)] and (b) a glassy carbon. (c–d) Peak analyses of (a–b) in a small range of energy-loss for the determination of sp^2 fraction. Note that the peaks at around 285 eV, 292 eV, and 300 eV represent the π^*-sp^2, σ^*-sp^3, and σ^*-sp^2 states, respectively. The light purple peak (at 287 eV) in (d) is a residual signal associated with the C-H σ^* state.

states. More detailed analysis is shown in Fig. 6.6c–d, where the EELS spectra are fitted by Gaussian peaks. The peaks at around 285 eV, 292 eV, and 300 eV represent the π^*-sp^2, σ^*-sp^3, and σ^*-sp^2 states, respectively. The light purple peak (at 287 eV) in (d) is a residual signal associated with the C-H σ^* state. Calculation using the intensity of these peaks indicates the sp^2 fraction is 77.2% in the 3D graphene,[1] where glassy carbon (100% sp^2) was used as the reference [Bernier et al. (2008); Brydson et al. (2008)]. A strong σ^*-sp^3 peak in the 3D graphene sample supports the presence of a remaining amorphous a-C carbon layer underneath the graphene layers.

6.2.4 Angle-Resolved Photoemission Spectroscopy

Measuring the electronic band structure and/or the density of states is useful to investigate the electrical properties of a material. There are several popular techniques for this study in the field of condensed matter physics. Scanning tunneling spectroscopy can be used to measure the density of states of a conductive material by obtaining the derivative of the tunneling current ($\frac{\mathrm{d}I}{\mathrm{d}V}$). This is because the tunneling current is the integration of the density of states with the Fermi–Dirac distribution over the electron energy (voltage), as described in section 4.2. However, this method cannot be used to measure the electronic band structure, as the momentum of the electrons also needs to be determined.

Angle-resolved photoemission spectroscopy (ARPES) is particularly useful for the study of low dimensional materials, because the photon is injected to a very thin layer on the surface of the material. Similar to most photoemission techniques, photons with sufficient energy strike the sample and knock out the electrons which carries the information of their energy in the material. In addition, the detector for ARPES can change its angle and find the direction that the emitted electrons move.

[1]The sp^2 fraction (x) is determined using a relation given by [Brydson et al. (2008)], $\frac{3x}{4-x} = \frac{I_\pi^U/I_\sigma^U}{I_\pi^R/I_\sigma^R}$, where I represents the intensity of the fitted Gaussian peak (for either π^* or σ^* states, denoted by the subscripts π and σ), and the superscripts U and R refer to the (unknown) sample and the reference.

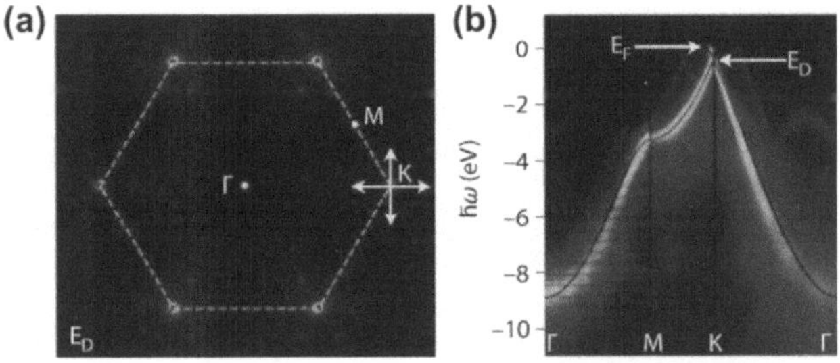

Figure 6.7 (a) A constant energy map of an ARPES measurement of a graphene on SiC, where the binding energy is close to the "Dirac" point. The dashed line illustrates the first Brillouin zone in the reciprocal lattice (Fig. 3.2). (b) The measured electronic band structure (energy vs. momentum) of graphene using ARPES, where the Fermi level and the "Dirac" point are clearly indicated. The solid black line shows the theoretical calculation. Reprinted with permission from [Bostwick et al. (2006)] Copyright Nature Publishing Group.

This would determine the momentum of the electrons, and hence plot the energy-vs.-momentum relation, i.e., the electronic band structure.

Figure 6.7 shows the ARPES measurement of a graphene on SiC [Bostwick et al. (2006)]. The first Brillouin zone is indicated by the dashed lines in Fig. 6.7a by mapping the constant energy near the "Dirac" points (the small circles shown at the corners of the hexagon). This is the same as that illustrated in Fig. 3.2. The measured electronic structure as shown in Fig. 6.7b is the same as the 2D version of the theoretical calculation demonstrated in Fig. 3.3. This indicates that ARPES is an effective and useful method for obtaining the information of the band structure of graphene. However, since this method requires ultra high vacuum, its application in graphene–electrolyte systems is limited. Therefore, it is important to find a technique which can be used in such systems.

6.2.5 Four-Probe Measurement

The first step to most experimental electrical investigations is often the measurement of resistance (or conductance), where the simplest method is to perform a two-probe measurement with a multimeter.

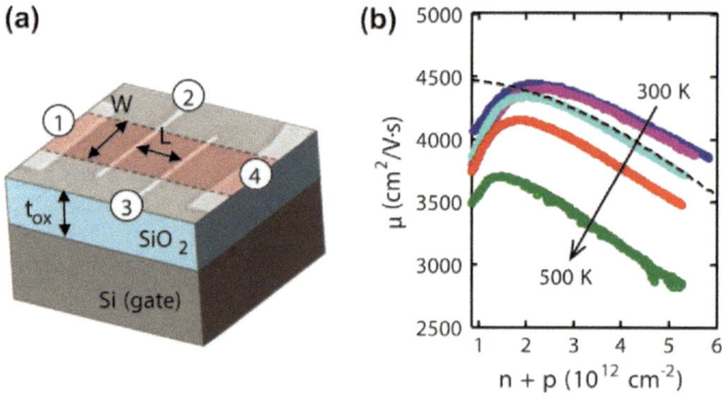

Figure 6.8 (a) Schematics of an experimental setup for four-probe measurement of graphene, where ①–④ are the electrical contacts and the red area enclosed by the dashed lines represents graphene. (b) The measured electron mobility as a function of the total carrier density at different temperature, where the dashed curve shows a theoretical fit. Reprinted with permission from [Dorgan et al. (2010)] Copyright American Institute of Physics.

This is useful for many daily tasks as this measurement is easy to take. However, it should be noted that the contact resistance between the probes of the multimeter and the sample is inevitably included.

Figure 6.8a shows the schematics of a graphene layer deposited with 4 well-defined electrical contacts (①–④). A regular two-probe measurement records the current (or voltage) between, for example, contacts ② and ③, while applying the voltage (or current) on the same electrodes. The electrical resistance (or conductance) can be derived using the Ohmic law, where the results include the resistance between the contacts and graphene. Note that the contact resistance is in series with the resistance that we are measuring. This indicates that the current flowing in the contacts is the same as that in the sample (graphene).

The commonly used four-probe measurement method avoids the problem of contact resistance by introducing two extra electrical contacts. In a four-probe measurement, the current is sent from contact ① to ④ while the voltage is measured between ② and ③, the potential drop between the sample and the contacts is not included due to the high impedance of the voltmeter. This indicates that the derived resistance will exclude the contribution from the contacts. Therefore, the four-probe measurement is believed more accurate than the two-probe setup.

It should be noted that, for the electrical measurement of a 2D material, we are often interested in the sheet resistance (or sheet conductance), rather than the regular resistance or resistivity for 3D materials. In general, the sheet resistance (ρ_s) of a 2D material is written as

$$\rho_s = R \cdot \frac{W}{L} = \frac{V}{I} \cdot \frac{W}{L}, \tag{6.3}$$

where $R = \frac{V}{I}$ is the measured resistance. V is the voltage and I is the current. As described earlier, $V = V_{23}$ and $I = I_{14} = I_{23}$ for the four-probe measurement. W and L are the width and length of the sheet between contacts ② and ③, as illustrated in Fig. 6.8a. In comparison with the resistivity of a 3D material defined as $\rho_v = R \cdot \frac{Wt}{L}$, where t is the thickness of the sample, t is not included in Eq. 6.3. This is because the thickness of a 2D material is not a primary concern, and it usually does not change too much between different samples made of the same type of material.

Note that the unit for resistivity is Ω·m, but the unit of sheet resistance should be the same as that of a regular resistance, as $\frac{W}{L}$ gives a dimensionless number in Eq. 6.3. It is also worth noting that in many literature the unit of the sheet resistance is written as Ω/□, in order to differentiate itself from that for the regular resistance.

The sheet resistance obtained by the four-probe measurement can be used to derive the mobility (or the charge carrier concentration) if the charge carrier concentration (or the mobility) is known. Figure 6.8b shows the electron mobility of a graphene sheet derived from the experimental results of a four-probe measurement, while the charge carrier concentration is calculated theoretically (section 3.2.1). The dashed line is an empirical fit which can be written as

$$\mu = \frac{\mu_0}{1 + (n/n_{\text{ref}})^{2.2}} \times \frac{1}{1 + (T/T_{\text{ref}} - 1)^3}, \tag{6.4}$$

where $\mu_0 = 4650\ \text{cm}^2\text{V}^{-1}\text{s}^{-1}$, $n_{\text{ref}} = 1.1 \times 10^{13}\ \text{cm}^2$, and $T_{\text{ref}} =$ 300 K. It can be seen from the figure and Eq. 6.4 that the mobility is dependent on the temperature and charge carrier concentration.

6.3 Classical Hall Effect of Graphene

Hall effect measurement is often performed to probe the electrical properties of a material such as charge carrier density, mobility, and magneto-electron transport, due to the simple experimental setup and the sophisticated theory. In principle, Hall effect measurements can be taken in vacuum, air, or liquid using a current (or voltage) source with a voltmeter. The Hall effect measurement is essentially an electrical measurement which detects different resistances of the studied device in the presence of magnetic field.

The Hall effect [Kittel (2004)] is the production of an electric field (Hall field, $\vec{E}_{\text{Hall}}$) in a conductor perpendicular to both current $\vec{j}$ and vertically applied magnetic field $\vec{B}$, as shown in Fig. 6.9a and b. The Hall effect measurement is one of the most commonly used methods to characterize the electronic properties of a material. As proposed in the previous section, Hall effect measurement of graphene in solution may provide another alternative for bio-sensing. In order to understand the experiments correctly, it is essential to study the carrier transport in the presence of magnetic fields, i.e. magnetotransport.

Magnetotransport in graphene can be described by either the Quantum Transport Theory [Gusynin and Sharapov (2005); Zheng and Ando (2002)] or the Boltzmann Transport Theory [Hwang et al. (2007b); Tiwari and Stroud (2009)], this depends on whether or not the Landau quantization should be taken into account. In the case of strong magnetic field, Hall conductivity of graphene displays an integer quantum Hall effect in the low-lying Landau levels [Gusynin and Sharapov (2005); Novoselov et al. (2005); Zhang et al. (2005); Zheng and Ando (2002)]. However, in the presence of a weak magnetic field, carrier transport in graphene can be described by the classical Boltzmann theory [Adam et al. (2007); Hwang et al. (2007b); Tiwari and Stroud (2009)].

If a Hall measurement is conducted in a weak magnetic field, a classical theory is adequate to model the carrier transport. In addition to the commonly used two-carrier model of graphene for solid devices [Hwang et al. (2007b); Tiwari and Stroud (2009)], a general expression of the resistance matrix of graphene in solution

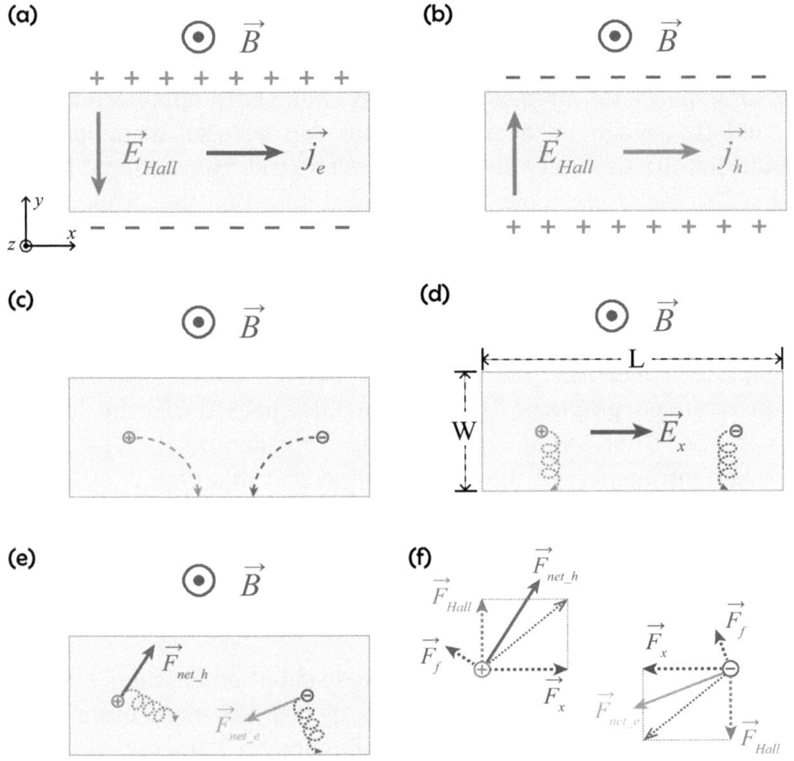

Figure 6.9 Hall effect in (a) an electron conductor and (b) a hole conductor, where $\vec{j}_e$ and $\vec{j}_h$ represent the electric current density. This indicates that the direction of the motion of electrons and holes in (a) and (b) is opposite to and the same as that of $\vec{j}_e$ and $\vec{j}_h$, respectively. The Drude model of carrier transport when the cyclotron radius is (c) comparable to, and (d) much less than the dimension of the conductor. (e) Drift of carriers taking into account all the forces. (f) Decomposition of the net forces, except the Lorentzian. W and L are the width and the length of the conductor.

is presented in this section, taking into account the difference in electron and hole mobility. This is because cations and anions of various concentrations are adsorbed onto graphene at different potentials [Bard and Faulkner (2000)], which have distinct effects on the scattering parameters of carriers in graphene. Moreover, the effects of electron-hole puddles induced by charged impurities on the transport parameters are discussed.

6.3.1 The Theory

It is important to review the theory of electron/hole transport in graphene in the presence of a perpendicularly applied magnetic field (i.e., magnetotransport), as this can be used to understand and predict the Hall effect measurement. The conventional theory used in many textbooks are developed based on the Drude model [Ashcroft and Mermin (1976); Kittel (2004)], where there is only one type of charge carriers (i.e., either electrons or holes). This is because electrons/holes are the dominant charge carriers in the conduction/valence band in the regular semiconductors with band gaps, and electrons are the charge carriers in metals. However, an undoped graphene is a semi-metal where the Fermi level is near the "Dirac" point, where the electrons and holes co-exist. The magnetotransport of the charge carriers in this case is described by the two-carrier model, as electrons and holes move in opposite directions in the presence of a magnetic field.

6.3.1.1 The Drude model

As described in section 2.1.1, the ion distribution function $f(\vec{r}, \vec{v}, t)$ is perceived as the particle density in position-momentum space (or phase space) in kinetic theory. Boltzmann Transport Equation (BTE, Eq. 2.1) describes the evolution of the particle distribution function in phase space. In general, it is impossible to obtain the global analytical solution without any assumption at this stage, even a recent study only presented the existence proof of the solution for long-range interactions [Gressman and Strain (2010)]. In order to obtain the useful solutions for a practical physics problem, it is common to perform integration of the BTE over different orders of momentum (appendix A).

In classical theory (both Drude and Sommerfeld models), the carriers in conductors are considered to be an electron (hole) gas [Ashcroft and Mermin (1976)]. Therefore the carrier transport can be described by BTE in this scheme. By deriving the 1st-order moment of BTE, the momentum conservation equation is obtained [Zhan and Hu (2011); Zhan et al. (2011a)]

$$m\frac{\mathrm{d}\vec{v}}{\mathrm{d}t} = q\left(\vec{E} + \vec{v} \times \vec{B}\right) - \frac{m\vec{v}}{\tau} - \frac{1}{n}\nabla\left(nk_{\mathrm{B}}T\right), \tag{6.5}$$

where m, $\vec{v}$, q, τ, n are the mass, velocity, charge, relaxation time and density of the carriers, respectively. k_B is Boltzmann's constant, $\vec{E}$, $\vec{B}$ and T are the electric field, magnetic field and temperature, respectively. Note this equation is the same as the Drude model when the carrier density gradient and the temperature gradient in the electron (hole) gas are not taken into account.

The carrier transport properties, such as resistance and Hall coefficient, are independent of time at equilibrium. Therefore, the left-hand side in Eq. 6.5 vanishes, and the equation becomes

$$0 = \vec{F}_{\text{net}} + q\vec{v} \times \vec{B} = \left(q\vec{E} - \frac{m\vec{v}}{\tau}\right) + q\vec{v} \times \vec{B}, \qquad (6.6)$$

where $\vec{F}_{\text{net}} = q\vec{E} - \frac{m\vec{v}}{\tau}$ is the net force applied on the carrier, excluding the Lorentzian. $\vec{F}_f = -\frac{m\vec{v}}{\tau}$ is the frictional force due to collision. This equation gives the **Drude model** for Hall effect.

When the cyclotron radius is comparable to the dimension of the device, an intuitive physical picture of the carrier transport is shown in Fig. 6.9c. The carriers move to one side in the conductor simply due to the Lorentz force, and therefore generate a Hall field if the density of electrons is different from that of holes. However, it is common in experiments that the dimension of the device could be much larger than that of the cyclotron radius. For example, the dimension of the Hall structure is around hundreds of microns, while the cyclotron radius is about 0.37 μm for $B = 0.1$ T, and $n = 10^{11}$ cm^{-2}, calculated by the following relation [Castro Neto et al. (2009)]

$$r_c = \frac{v_F}{\omega_c} = \frac{v_F m^*}{qB} = \frac{\hbar\sqrt{\pi n}}{qB}, \qquad (6.7)$$

where $v_F \approx 10^6$ m/s is the Fermi velocity [Wallace (1947)], $\omega_c = \frac{qB}{m^*}$ is the cyclotron frequency, and $m^* = \frac{\hbar\sqrt{\pi n}}{v_F}$ is the cyclotron mass for carrier (electron/hole) in graphene. In this case, the carrier would complete its *full* cycle of motion in the magnetic field before it hits the side wall of the conductor.

However, a careful study to Eq. 6.6 indicates that the drift velocity of both electrons and holes is in the direction of $-\hat{y}$, if $\vec{B} = B\hat{z}$, and $\vec{E} = \vec{E}_x = E\hat{x}$, where $\hat{x}$, $\hat{y}$, and $\hat{z}$ are the unit vectors. Rearranging Eq. 6.6 and multiplying $\vec{B}$ on both sides gives the drift velocity

$$\vec{v} = \frac{\vec{F}_{\text{net}} \times \vec{B}}{qB^2}. \qquad (6.8)$$

If $\vec{F}_{\text{net}} = q\vec{E}_x$, the drift velocity, $\vec{v} = \frac{\vec{E}_x \times \vec{B}}{B^2}$, is in the same direction and of the same magnitude for both electrons and holes, as illustratively demonstrated in Fig. 6.9d.

As also shown in Fig. 6.9c, the drift velocity of electrons and holes are always in the same direction when we consider only $\vec{E}_x$ in the conductor, regardless the magnitude of cyclotron radius. This leads to the generation of the Hall fields in the opposite direction for an electron/hole conductor.

In general, a few forces are applied on the carriers, such as the electrostatic force due to $\vec{E}_x$ and $\vec{E}_{\text{Hall}}$, the frictional force $\vec{F}_f$ due to collisions, and so on. This will change the drift velocity, and hence affect the resistance and the Hall coefficient. Figure 6.9e shows the corresponding physical picture. Note that the direction of $\vec{F}_{\text{net}}$ for electrons and holes are not parallel to each other. This is because $\vec{F}_f$ must in the opposite direction of drift velocity, i.e., perpendicular to $\vec{F}_{\text{net}} = \vec{F}_x + \vec{F}_{\text{Hall}} + \vec{F}_f$, as indicated in Fig. 6.9f.

6.3.1.2 Cyclotron mass

The concept of the cyclotron mass was mentioned earlier in Eq. 6.7. However, the definition and the derivation process have not been given. Since the electrons in graphene are considered as massless Fermions due to the gapless energy bands (Chapter 3), which lead to the ultrahigh mobility, it is important to understand the origin of the cyclotron mass in graphene.

In classical physics, charged particles circles in a plane which is perpendicular to the applied magnetic field in real space, due to the Lorentz force $\vec{F} = q\vec{v} \times \vec{B}$. q and $\vec{v}$ are the charge and the velocity of the charged particle, and $\vec{B}$ represents the magnetic flux density (SI unit: Tesla, T). As $\vec{F}$ is perpendicular to $\vec{v}$, the Lorentz force has no contribution to energy. However, as discussed in Chapter 2, it is more accurate and convenient to describe the motion and energy of electrons in crystals using electronic band structure in solid state physics. Therefore, it is necessary to translate the Lorentz force from classical physics to the modern theory of graphene.

Figure 6.10a illustrates the electronic band structure of graphene near the "Dirac" point. When there is a magnetic field, the electron circulates in a iso-energy plane in the k-$\mathcal{E}$ space as the Lorentz force

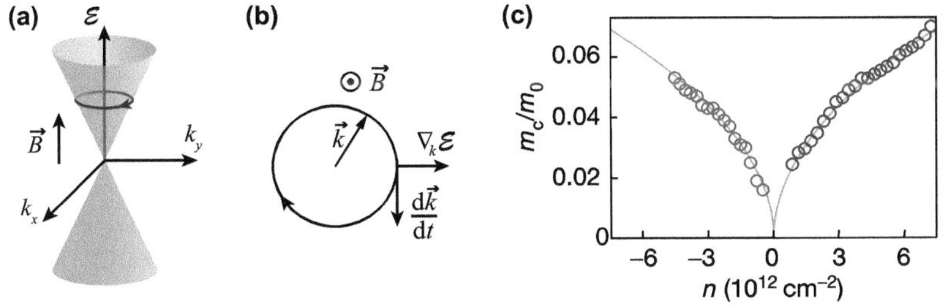

Figure 6.10 (a) Electronic band structure near the "Dirac" point, where the black ring represents the trajectory of an electron moving in the k-$\mathcal{E}$ space in the presence of a magnetic field. (b) Top view of the electron trajectory in the k-$\mathcal{E}$ space. (c) Measured cyclotron mass of charge carriers in graphene (open circles) fitted by Eq. 6.12 (solid curves), where m_0 is the mass of a free electron, and the positive and the negative n corresponds to electrons and holes, respectively. Reprinted with permission from [Novoselov et al. (2005)] Copyright Nature Publishing Group.

does not change the energy, as demonstrated by the black ring. Note that the magnetic field is applied in the z-direction, while k_x and k_y indicate that the graphene is placed on a xy plane. Since the direction of k_x and k_y is the same as that of the x and y axes in real space, electrons can circle in the k-plane similar to the way that it does in a Cartesian plane. In comparison with the Lorentz force in real space, where $m\frac{\mathrm{d}\vec{v}}{\mathrm{d}t} = q\vec{v} \times \vec{B}$, the Lorentz force in the k-$\mathcal{E}$ space can be written as

$$\hbar\frac{\mathrm{d}\vec{k}}{\mathrm{d}t} = \frac{q}{\hbar}\nabla_k\mathcal{E} \times \vec{B}. \tag{6.9}$$

Note that the relation $m\vec{v} = \hbar\vec{k}$ is used when deriving Eq. 6.9. $\hbar = \frac{h}{2\pi}$, h is the Planck's constant, and $\vec{k}$ is the wave vector of electrons. $\nabla_k\mathcal{E}$ is the gradient of energy in k-plane, which also represents the group velocity of the electrons in real space.

The top view of the trajectory of electrons in the k-$\mathcal{E}$ space is shown in Fig. 6.10b. Note that the radius of the circle is $\vec{k}$, as the dispersion relation of electrons in graphene is linear near the "Dirac" point (Eq. 3.4). The cyclotron frequency (ω_c) is defined as the angular frequency of the cyclotron of electrons, which could be expressed as $\omega_c = \frac{|\vec{v}|}{r}$ in real space, where $|\cdots|$ represents the magnitude of a vector (i.e., the norm). Hence, for a circle shown in

Fig. 6.10b, the cyclotron frequency can be written using $\vec{k}$ as

$$\omega_c = \frac{1}{k}\left|\frac{d\vec{k}}{dt}\right|, \tag{6.10}$$

where the analogy $k \sim r$ and $\frac{d\vec{k}}{dt} \sim \frac{d\vec{r}}{dt} = \vec{v}$ are used.

Substitution of Eqs. 6.9 and 6.10 into the classical definition of the **cyclotron mass** where $m^* = \frac{qB}{\omega_c}$, one obtains

$$m^* = \frac{qBk}{dk/dt} = \hbar^2 k \frac{\partial k}{\partial \mathcal{E}}. \tag{6.11}$$

Recall that the dispersion relation of electrons in graphene near the "Dirac" point is $\mathcal{E} = \hbar v_F k$, where $v_F \approx 10^6$ m/s is the Fermi velocity, Eq. 6.11 can be written as

$$m^* = \frac{\hbar k_F}{v_F} = \frac{\hbar\sqrt{\pi n}}{v_F}, \tag{6.12}$$

where n represents the carrier density. A relation $n = g\frac{\pi k_F^2}{4\pi^2} = \frac{k_F^2}{\pi}$ is used. $g = 4$ is the degeneracy which has taken into account the spin and the valley degeneracy. k_F is the wave number (or momentum) at the Fermi level, as this energy level is important for charge carrier transport.

Figure 6.10c shows the measured cyclotron mass (m_c) of holes (negative n) and electrons (positive n) in graphene by investigating the SdH oscillations at different temperature [Novoselov et al. (2005)], where m_0 is the free electron mass. The solid curves are the theoretical fitting to m_c using Eq. 6.12, where the square root dependence of the cyclotron mass on the charge carrier density is confirmed.

6.3.1.3 The two-carrier model

As can be seen from Fig. 6.9, both $\vec{j}$ and $\vec{E}$ have components in $\hat{x}$ and $\hat{y}$ directions, which are dependent of each other. Therefore, the simple Ohmic law results in a conductivity tensor for carrier transport in graphene in the presence of magnetic field, i.e. $\vec{j} = \overleftrightarrow{\sigma}\vec{E}$ where $\overleftrightarrow{\sigma}$ is the conductivity tensor (matrix). Rearranging Eq. 6.6

gives $\vec{v} = \mu(\vec{E} + \vec{v} \times \vec{B})$, where $\mu = \frac{q\tau}{m}$ is the mobility. Hence,

$$\begin{pmatrix} j_x \\ j_y \end{pmatrix} = \frac{qn\mu}{1+(\omega_c\tau)^2}\begin{pmatrix} 1 & \omega_c\tau \\ -\omega_c\tau & 1 \end{pmatrix} \cdot \begin{pmatrix} E_x \\ E_y \end{pmatrix} = \sigma \begin{pmatrix} E_x \\ E_y \end{pmatrix}, \quad (6.13)$$

where the relations $\vec{j} = qn\vec{v}$ and $\mu B = \omega_c\tau$ have been used.

Since the total current is contributed to by both electrons and holes, $\vec{j}_t = \vec{j}_e + \vec{j}_h$. Given $q_h = -q_e = e$, the total conductance ($\overleftrightarrow{\sigma}_t = \sigma_e + \sigma_h$) is written

$$\begin{aligned} \overleftrightarrow{\sigma}_t =& \frac{en_h\mu_h}{1+(\mu_h B)^2}\begin{pmatrix} 1 & \mu_h B \\ -\mu_h B & 1 \end{pmatrix} \\ &- \frac{en_e\mu_e}{1+(\mu_e B)^2}\begin{pmatrix} 1 & \mu_e B \\ -\mu_e B & 1 \end{pmatrix}. \end{aligned} \quad (6.14)$$

The *total resistance matrix* can be obtained by finding the inverse of $\overleftrightarrow{\sigma}_t$,

$$\begin{aligned} \overleftrightarrow{\rho}_t = R_0 &\begin{pmatrix} n_h\mu_h[1+(\mu_e B)^2] - n_e\mu_e[1+(\mu_h B)^2] \\ B\{n_h\mu_h^2[1+(\mu_e B)^2] - n_e\mu_e^2[1+(\mu_h B)^2]\} \end{pmatrix. \\ &\quad \begin{matrix} B\{n_e\mu_e^2[1+(\mu_h B)^2] - n_h\mu_h^2[1+(\mu_e B)^2]\} \\ n_h\mu_h[1+(\mu_e B)^2] - n_e\mu_e[1+(\mu_h B)^2] \end{matrix}\Bigg) \quad (6.15) \\ =& \begin{pmatrix} \rho_{xx} & \rho_{xy} \\ \rho_{yx} & \rho_{yy} \end{pmatrix}, \end{aligned}$$

where $R_0 = \frac{1}{e[(\mu_e\mu_h B)^2(n_e-n_h)^2+(n_e\mu_e-n_h\mu_h)^2]}$, $\rho_{xx} = \rho_{yy}$, and $\rho_{yx} = -\rho_{xy}$. This is the general form of the **two-carrier model** of graphene. Since $j_y = 0$ at equilibrium in Hall measurement, $E_x = \rho_{xx}j_x + \rho_{xy}j_y = \rho_{xx}j_x$, similarly, $E_y = \rho_{yx}j_x$. It should be noted that $E_y = E_{\text{Hall}}$ as indicated in Fig. 6.9. In Hall measurements of graphene, it is of interest to study the sheet density instead of the volumetric; i.e. $n = \frac{N}{W \cdot L}$, where N is the total number of carriers, W and L are the width and length of the conducting channel as shown in Fig. 6.9. Then the 2D current density can be written $j_x = e\frac{N}{W \cdot L}\frac{L}{t} = \frac{I}{W}$. Hence, $\rho_{xx} = \frac{E_x}{j_x} = \frac{V_x}{I_x} \cdot \frac{W}{L}$ and $\rho_{yx} = \frac{E_y}{j_x} = \frac{V_{\text{Hall}}}{I_x} \cdot \frac{W}{W} = \frac{V_{\text{Hall}}}{I_x}$, where ρ_{xx} and $-\frac{\rho_{yx}}{B}$ represent the sheet and the Hall resistances, respectively. The **Hall resistance** defined as $R_{\text{H}} = -\frac{V_{\text{Hall}}}{I \cdot B}$ is useful in the experiments, as the charge carrier density and mobility can be extracted (section 6.3.2).

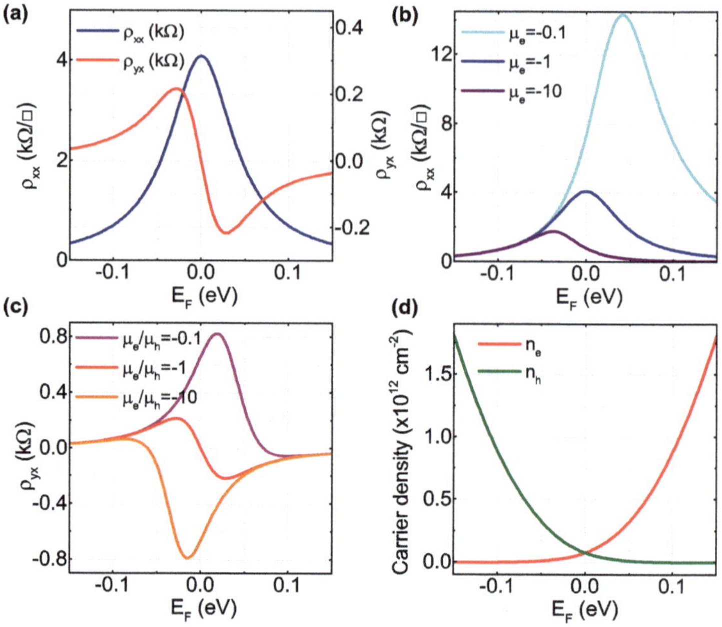

Figure 6.11 (a) Resistance tensor elements ρ_{xx} and ρ_{yx} as a function of the gate voltage ($-\mu_e = \mu_h = 1$). (b–c) Changes of ρ_{xx} and ρ_{yx} with μ_e (from -0.1 to -10 $\mathrm{m}^2/(\mathrm{Vs})$), respectively. Note $\mu_h = 1$ $\mathrm{m}^2/(\mathrm{Vs})$, and $B = 0.1$ T. (d) Carrier density distribution as a function of Fermi level.

Figure 6.11a plots ρ_{xx} and ρ_{yx} when $-\mu_e = \mu_h = 1(m^2/(Vs))$ and $B = 0.1T$. It can be seen that ρ_{xx} and ρ_{yx} are symmetric and antisymmetric with respect to the "Dirac" point ($E_D \equiv 0$), respectively. The peak in ρ_{xx} normally indicates the "Dirac" point, as it usually represents the transition from the hole region ($E_\mathrm{F} < 0$) to the electron region ($E_\mathrm{F} > 0$).

If graphene is chemically doped [Lherbier et al. (2008)], or if there is charged impurity in the environment close to the graphene [Adam et al. (2007)], it is possible to have $\mu_h \neq -\mu_e$, as briefly discussed in section 5.1.1. As a result, the symmetries in the transport properties can change drastically, as shown in Fig. 6.11b and c. It can also be seen from Fig. 6.11b that the magnitude of ρ_{xx} decreases as $|\mu_e|$ increases, which is expected. More interestingly, the antisymmetric shape of the ρ_{yx} curve becomes a peak and a valley when $-\mu_e < \mu_h$ and $-\mu_e > \mu_h$, respectively. Note that V_G is calculated using an aforementioned geometry-related factor a, and its magnitude changes with different devices.

It is then convenient to rewrite the sheet and the Hall resistances by dividing $(n_h\mu_e)^2$ from both the numerators and the denominators, hence

$$\rho_{xx} = \frac{\frac{1}{n_h\mu_e}\left\{\frac{\mu_h}{\mu_e}\left[1+(\mu_e B)^2\right] - \frac{n_e}{n_h}\left[1+(\mu_h B)^2\right]\right\}}{e\left[(\mu_h B)^2\left(\frac{n_e}{n_h}-1\right)^2 + \left(\frac{n_e}{n_h}-\frac{\mu_h}{\mu_e}\right)^2\right]}, \tag{6.16a}$$

$$\rho_{yx} = \frac{\frac{B}{n_h}\left\{\left(\frac{\mu_h}{\mu_e}\right)^2\left[1+(\mu_e B)^2\right] - \frac{n_e}{n_h}\left[1+(\mu_h B)^2\right]\right\}}{e\left[(\mu_h B)^2\left(\frac{n_e}{n_h}-1\right)^2 + \left(\frac{n_e}{n_h}-\frac{\mu_h}{\mu_e}\right)^2\right]}. \tag{6.16b}$$

This indicates that ρ_{xx} and ρ_{yx} are strongly dependent on the ratio between μ_e and μ_h.

As observed in Fig. 6.11b and d, although the peak of ρ_{xx} moves from right to left as $|\mu_e|$ increases, the Fermi level remains at the "Dirac" point, i.e. $E_F = E_D = 0$ when $V_G = 0$. This is because the difference in the electron/hole mobility does not have effects on the charge carrier density, i.e. Eq. 3.6. It can be shown by numerical methods that the calculated carrier densities are exactly the same as those illustrated in Figs. 6.11d and 3.5b. In other words, the shift of the peak in ρ_{xx} does not necessarily indicate the doping of graphene.

It is expected from Eq. 6.16 that changing the value of μ_h while fixing μ_e will have similar (but not exactly the same) effects on the transport properties, as $\frac{\mu_h}{\mu_e}$ is similar. These changes will not be shown here, readers who are interested in these results are encouraged to plot the new relations.

In the case where graphene is not immersed in solution (for example, a graphene/SiO_2/Si device in vacuum), the scattering events for electrons and holes are similar. This is because the charged impurities in the solid environment follow Gaussian distribution which is symmetric with respect to "Dirac" point [Martin et al. (2008)], and they are immobile in response to the varying gate voltage. Therefore a common approximation $\mu_h \approx -\mu_e = \mu$ is used, and Eq. 6.15 can be simplified,

$$\overleftrightarrow{\rho_t} = \frac{1+(\omega_c\tau)^2}{e\mu[n_t^2+(\omega_c\tau)^2 n_{\text{net}}^2]}\begin{pmatrix} n_t & -\omega_c\tau n_{\text{net}} \\ \omega_c\tau n_{\text{net}} & n_t \end{pmatrix}. \quad (6.17)$$

This equation describes the two-carrier model of graphene for symmetric electrons/holes [Hwang et al. (2007b); Tiwari and Stroud (2009)], and it is often used to describe the graphene devices that do not involve solution [Novoselov et al. (2005, 2004); Zhang et al. (2005)]. It should be noted that the symmetries of ρ_{xx} and ρ_{yx} derived from an ideal model, as shown in Fig. 6.11a, can still be observed in a relatively dirty graphene/SiO_2/Si sample. This is because the electron-hole puddles in the dirty graphene sample follow a Gaussian distribution [Martin et al. (2008)].

For conventional semiconductors where $n_t = |n_{\text{net}}|$ (because $|n_h + n_e| = |n_h - n_e|$ when there is one type of charge carriers), Eq. 6.17 can be reduced

$$\overleftrightarrow{\rho_t} = \frac{1}{e\mu n_t}\begin{pmatrix} 1 & -\mu B \\ \mu B & 1 \end{pmatrix}. \quad (6.18)$$

Note that this is also true for other two-carrier materials where the mean free path (λ) of the carrier is much shorter than the cyclotron radius (r_c), $\lambda \ll r_c$ (i.e. $\omega_c\tau \ll 1$) when V_G is away from 0V, as $n_t \approx n_{\text{net}}$ at high gate voltage. This result is the same as that of the Drude model which can be inferred from section 6.3.1.1.

6.3.2 Experimental Setup

As described in Fig. 6.9 and section 6.3.1.3, Hall effect measurement is used to characterize the electrical properties of the material through the investigation of the magnetotransport. It is therefore useful to measure the sheet and Hall resistances, in order to obtain the total resistance matrix (Eq. 6.15). This result can then be used to directly evaluate the charge carrier concentration and the mobility for conventional semiconductors (Eq. 6.18).

6.3.2.1 Hall structure

In essence, the detection of the Hall effect is the measurement of the sheet and Hall resistances. Figure 6.12 shows a commonly used

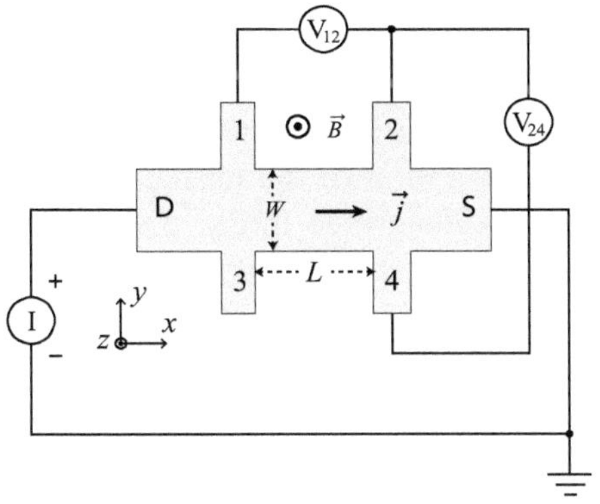

Figure 6.12 Schematics of a Hall bar with the circuit used to measure the sheet and Hall resistances.

structure (**Hall bar**) in many Hall effect measurements. D and S represent drain and source for electrons in the Hall bar, which are the conventional names for field effect transistors (sections 2.3.3 and 3.3.1). Four electrical contacts (1–4) on the side are defined to measure the voltage along and across the D-S direction within the structure. A constant current source (I) is connected to D and S when a magnetic field ($\vec{B}$) is applied perpendicularly.

The voltage between electrical contacts 1 and 2 (V_{12}) is measured to obtain the sheet resistance. This configuration is the same as that for the four-probe measurements (section 6.2.5). The sheet resistance is then expressed as

$$\rho_{xx} = \frac{V_{12}}{I} \cdot \frac{W}{L}, \tag{6.19}$$

where W and L are the width and length of the conducting channel between contacts 1, 3 and 2, 4 of the Hall bar.

V_{24} measured between contacts 2 and 4 for the configuration in Fig. 6.12 is used to derive the Hall resistance

$$R_{\mathrm{H}} = -\frac{\rho_{yx}}{B} = -\frac{V_{24}}{IB}. \tag{6.20}$$

For conventional semiconductors, the charge carrier concentration and mobility can be estimated using the following equations

$$n = \frac{1}{eR_{\mathrm{H}}}, \tag{6.21a}$$

$$\mu = \frac{1}{en\rho_{xx}} = \frac{R_{\rm H}}{\rho_{xx}}, \tag{6.21b}$$

where $e = 1.60217662 \times 10^{-19}$ C is the elementary charge. Note that one must pay attention to the direction of V_{24} and $\vec{B}$ during the measurement. As illustrated in Fig. 6.9a and b, V_{24} is positive and negative when the electrons and holes are the majority charge carriers, respectively. This indicates that $R_{\rm H}$ and n are negative for electrons and positive for holes, because of the negative sign in Eq. 6.20. If the direction of the magnetic field is reversed, V_{42} should be measured instead of V_{24}.

It should be noted that Eq. 6.21 can only be used to characterize the electrical properties of graphene when the Fermi level is far enough from the "Dirac" point (i.e., high gate voltage), where electrons and holes are the majority carriers, as can be inferred from Fig. 6.11. When the gate voltage is low, Eq. 6.16 should be adopted to analyze the measured sheet and Hall resistances of graphene.

6.3.2.2 Van der Pauw method

The well-defined electrical contacts in the Hall bar enables the accurate measurement of the sheet and Hall resistances, however, it could be inconvenient or even impossible in some cases to fabricate such a structure. Hence, it is necessary to develop a technique which can be used to perform Hall effect measurement in a sample with an arbitrary geometry. **Van der Pauw method** provides a mathematical approach to analyze the measured current in an arbitrary shaped 2D material.

Figure 6.13 shows a simple example of the van der Pauw method where the 2D material has a rectangular shape. Four electrical contacts (1–4) are defined around the perimeter of the sample (in this case, at the corners). A constant current source (I) is connected to contacts 3 and 4, while the voltage is measured between 1 and 2 (V_{12}), as illustrated in Fig. 6.13a. It is then necessary to understand the relation between V_{12} and I.

In an ideal (semi-infinite) model, the rectangular geometry is considered large enough such that the current flow out from a point spread away radially. The dashed curve in Fig. 6.13b shows the ideal model of the current from contact 3. The current density is then

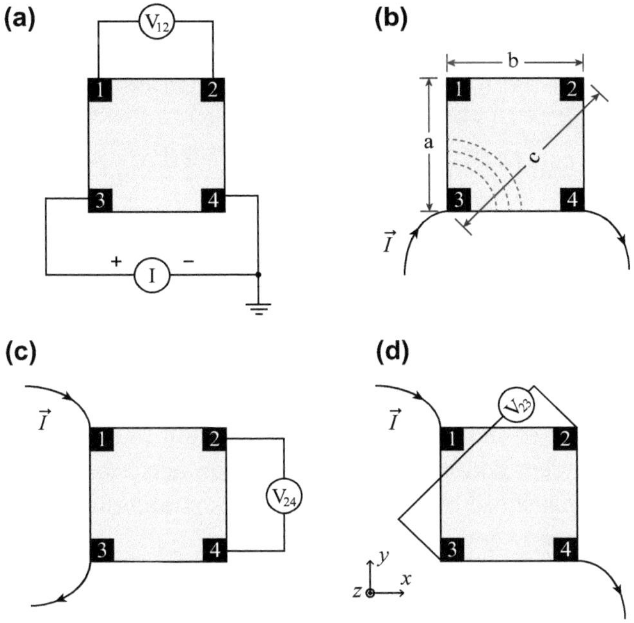

Figure 6.13 Schematics of the van der Pauw method for the measurement of Hall effect, where the electrical contacts (1–4) are defined at the corners of a rectangular shaped 2D material. (a) and (c) Configurations for measuring the sheet resistance, where a constant current source is connected to two of the contacts while the voltage is measured on the opposite two. (b) Analysis of the current in the material. (d) Measurement of the Hall coefficient, where the magnetic field is applied perpendicularly on the material.

$j = \frac{2I}{\pi r}$, where r is the radius of the dashed curve. Therefore, the electric field between a point on the curve and contact 3 is

$$E(r) = \frac{j}{\sigma_s} = j\rho_s = \frac{2I\rho_s}{\pi r}, \tag{6.22}$$

where σ_s and ρ_s are the sheet conductance and the sheet resistance of the material. Hence, the voltage between 1 and 2 can be calculated

$$V_{12,3} = -\int_{r_2}^{r_1} E\,\mathrm{d}r = \int_{r_3}^{r_2} E\,\mathrm{d}r - \int_{r_3}^{r_1} E\,\mathrm{d}r = \frac{2I\rho_s}{\pi}\ln\left(\frac{c}{a}\right), \tag{6.23}$$

where c is the distance between contacts 2 and 3 (the diagonal length of the rectangle), and a is the distance between 1 and 3 (a side length of the rectangle). It should be noted that, due to the finite shape of the sample structure, the influence of contact 4 should be taken into account. Similar to the derivation process of Eq. 6.23, the voltage between 1 and 2 can be written as

$$V_{12,4} = -\left(\int_{r_4}^{r_2} E\,\mathrm{d}r - \int_{r_4}^{r_1} E\,\mathrm{d}r\right) = -\frac{2I\rho_s}{\pi}\ln\left(\frac{a}{c}\right). \qquad (6.24)$$

Note that the negative sign is the result of the direction of the current in contact 4 (opposite to that in 3). The measured voltage between contacts 1 and 2 is then the average of $V_{12,3}$ and $V_{12,4}$,

$$V_{12} = \frac{V_{12,3} + V_{12,4}}{2} = \frac{I\rho_s}{\pi}\ln\left(\frac{c^2}{a^2}\right). \qquad (6.25)$$

Figure 6.13c shows a similar measurement to Fig. 6.13a, except that the current source is connected to contacts 1 and 3, while the voltage is measured between 2 and 4. It is straightforward to write the relation between V_{24} and I as

$$V_{24} = \frac{I\rho_s}{\pi}\ln\left(\frac{c^2}{b^2}\right), \qquad (6.26)$$

where b is the distance between contacts 1 and 2 (the other side length of the rectangle).

As $\frac{a^2+b^2}{c^2} = 1$, Eqs. 6.25 and 6.26 can be re-arranged as

$$\exp\left(-\frac{\pi R_{12}}{\rho_s}\right) + \exp\left(-\frac{\pi R_{24}}{\rho_s}\right) = 1, \qquad (6.27)$$

where $R_{12} = \frac{V_{12}}{I}$ and $R_{24} = \frac{V_{24}}{I}$ are the measured "effective" resistance. Equation 6.27 can be used to derive the sheet resistance ρ_s for the van der Pauw method. For example, if $a = b$, $R_{12} = R_{24}$. Hence, the sheet resistance can be obtained as

$$\rho_s = \frac{\pi R_{12}}{\ln 2}. \qquad (6.28)$$

The Hall field is measured using the configuration illustrated in Fig. 6.13d. As discussed in section 6.3.1.3, the measured Hall resistance is not dependent on the geometry. However, the Hall field must be measured in the direction which is perpendicular to the direction of the electric current on the plane of the 2D material. Due to the arbitrary shape of the sample, the diagonal between contacts 2 and 3 is not necessarily perpendicular to that between 1 and 4. The measured Hall field then includes the contribution to the electric current from both x and y direction, as $\vec{E} = \overleftrightarrow{\rho_t} \cdot \vec{j}$. Note that the definition of the Cartesian coordinates in Fig. 6.13 is different from that in Fig. 6.9. It is then necessary to remove the contribution of the sheet resistance to the Hall resistance.

As can be inferred from Eq. 6.6, $\vec{v} = \mu(\vec{E} + \vec{v} \times \vec{B})$. Hence, $\vec{j} = qn\mu(\vec{E} + \vec{v} \times \vec{B})$, which can also be written as

$$j_x = \sigma_s E_x + \mu j_y B_z, \tag{6.29a}$$

$$j_y = \sigma_s E_y - \mu j_x B_z, \tag{6.29b}$$

where $\vec{B} = B_z \hat{e}_z$ and $\sigma_s = qn\mu$ is the sheet conductance. Rearrangement of Eq. 6.29 gives

$$E_x = \rho_s j_x - \mu \rho_s j_y B_z, \tag{6.30a}$$

$$E_y = \rho_s j_y + \mu \rho_s j_x B_z. \tag{6.30b}$$

The voltage between contacts 2 and 3 is then

$$V_{23,+} = -\int_{r_3}^{r_2} (E_x \mathrm{d}x + E_y \mathrm{d}y). \tag{6.31}$$

When the magnetic field is reversed, the measured voltage is recorded as $V_{23,-}$. If a voltage V_H is defined as

$$V_\mathrm{H} = \frac{V_{23,+} - V_{23,-}}{2} = \mu \rho_s B_z \left(\int_{x_3}^{x_2} j_y \mathrm{d}x - \int_{y_3}^{y_2} j_x \mathrm{d}y \right), \tag{6.32}$$

this then represents the Hall voltage. In the case of a 3D material, the electric current (I) is the integration of the current density and the cross-sectional area. When the material is 2D, $I = \int \vec{j} \cdot \hat{n} \mathrm{d}l$. For the van der Pauw structure shown in Fig. 6.13d, l represents the line connecting contacts 2 and 3, and $\hat{n}$ is normal to l within the plane. Therefore, the Hall voltage (i.e., the Hall resistance) can be obtained

$$V_\mathrm{H} = \mu \rho_s B_z \int_{r_3}^{r_2} \vec{j} \cdot \hat{n} \mathrm{d}l = \mu \rho_s B_z I. \tag{6.33}$$

This indicates that the van der Pauw method can also be used to perform the Hall effect measurement. However, it should be noted that, it is essential to ensure the inversed magnetic field to remain the same magnitude as the previous one. In addition, although a rectangular shaped 2D material is used in the derivation process in the section, the discussion of the van der Pauw method using an arbitrary geometry can be found in [Lundstrom (2009)].

Chapter 7

Experimental Methods for Electrolyte

Experimental studies of electrical processes in electrolyte at the electrode–electrolyte interface are usually performed with electrochemical methods. These techniques can be used to investigate the electron transfer process and/or the ion structure at the electrode–electrolyte interface, as discussed in Chapter 4. In essence, the electrochemical methods detects the response of an electrode–electrolyte system by sending an (input) electrical signal. Thanks to modern electronics, a range of electrochemical techniques are developed for various purposes, due to the numerous choices of the waveform of the input electrical signal. Note that this chapter is by no means a thorough review of electrochemical methods; it rather serves as introductory guide of the techniques which are relevant to probing the electronic properties in electrode immersed in electrolyte. Here we mainly focus on the working principle of these methods, where the empirical models are selectively presented.

7.1 Cyclic Voltammetry

Cyclic voltammetry (CV) is one of the most common electrochemical methods used to identify the redox reactions and/or calculate the interfacial capacitance. This is suitable for a range of different

Graphene–Electrolyte Interfaces: Electronic Properties and Applications
Hualin Zhan

ISBN 978-981-4774-94-9 (Hardcover), 978-1-003-04487-1 (eBook)
www.jennystanford.com

applications, such as molecular detection, electrochemical corrosion (anodizing aluminum and produce ordered nanoporous alumina), electroplating, determining electrochemical capacitance, and so on.

As described in section 4.2, the charge transfer process between the electrode and electrolyte is enhanced significantly when the Fermi level of the electrode is close to the electrochemical potential of the electrolyte. The Fermi level can be tuned by applying a potential across the electrode–electrolyte interface. This potential can be applied by employing two electrodes (i.e., working electrode and counter electrode) in the electrolyte. However, due to the intrinsic redox potential of the material of the counter electrode, it is difficult to find the actual potential on the working electrode. Therefore, it is common to introduce an additional electrode with a fixed redox potential to the experimental setup. This electrode is called reference electrode, and such a system is defined as three-electrode electrochemical cell. An internationally accepted reference electrode is the Standard Hydrogen Electrode (or Normal Hydrogen Electrode) formed by blowing hydrogen gas to the platinum electrode in the electrolyte. Due to the convenience in electrode fabrication, another commonly used reference electrode is silver-silver chloride (Ag/AgCl) electrode [Bard and Faulkner (2000)].

Figure 7.1a is the schematic of a three-electrode electrochemical cell, where the voltage is applied between the working (WE) and the counter (CE) electrodes. The reference electrode (RE) placed closed to WE is used to measure the electrochemical potential of WE, as the electrochemical potential on RE is known. This idea is the same to four-probe measurement in electrics where the contact resistance is excluded from the results. The equivalent circuit of such a setup is shown in Fig. 7.1b.

In regards to CV, the voltage is linearly scanned back and forth between certain voltages. Since this changes the energy level of the electrons in electrode with respect to the electrochemical potential of the electrolyte, charge transfer processes occur differently. The obtained current can be described using Eqs. 4.28–4.31. As shown in Fig. 7.1d, when the Fermi level of the electrode is higher than the electrochemical potential of the species in electrolyte, i.e. the

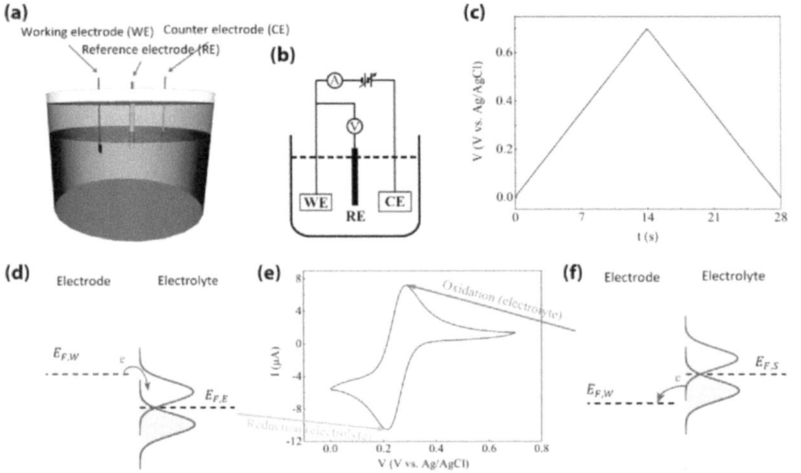

Figure 7.1 (a) Schematic of a three-electrode electrochemical system, where the working electrode (WE), counter electrode (CE), and reference electrode (RE) are indicated. (b) The equivalent circuit of (a). (c) Voltage measured between WE and RE as a function of time for a Cyclic Voltammetry (CV) measurement at the scanning rate of 100 mV/s. (d,f) Representation of reduction and oxidation processes for species in electrolyte when the potentials on the electrode are relatively more negative and positive, respectively. (e) Cyclic voltammogram obtained in $Fe(CN)_6^{-4}$, where a platinum electrode and a Ag/AgCl (1 M) electrode are used as WE and RE, respectively. The scanning voltage is shown in (c).

electrode potential is more negative, electrons move from electrode to electrolyte. The species in the electrolyte gains electrons, therefore, the species is reduced. Similarly, if the electrode potential is more positive, the species is oxidized, as illustrated in Fig. 7.1f. These processes are reflected by the two peaks in cyclic voltammogram, as indicated in Fig. 7.1e. This is because the charge transfer processes introduce current changes. Therefore, by distinguishing the redox peaks for the species in cyclic voltammogram, we are able to detect different molecules.

A question comes naturally when correlating Fig. 7.1e–d and f is that why the redox reactions manifest themselves as peaks in the voltammogram, i.e., the absolute value of the current should not reduce when the applied voltage (V) is greater/smaller than the oxidation/reduction potential. This is because the reactants can be depleted at the electrode–electrolyte interface. Therefore, the detailed study of CV involves the mass transfer process (which can be described using the fluid mechanics discussed in section 4.1.1), readers who are interested in this topic are encouraged to consult [Bard and Faulkner (2000)]. Popular methods used in voltammetry research to engineer the depletion zone include introducing microelectrodes or/and hydrodynamic techniques (microfluidics, rotating disk, etc.). Note that useful information about the chemical reaction and the electrolyte can be extracted from the analysis of the cyclic voltammogram. For example, if the potential difference between the reduction and oxidation peaks is around $\frac{57}{n}$ mV at 25°C, the reaction is considered electrochemically reversible. In addition, the diffusion coefficient (D) for the reactants in solution can be calculated using the following equation,

$$I = 0.4463nFA\left(\frac{nFDv}{R_G T}\right)^{\frac{1}{2}} c, \tag{7.1}$$

where n is the number of electrons involved in a single reaction, $F = eN_A = 96485.33289\ \text{C}\cdot\text{mol}^{-1}$ is the Faraday constant, N_A is the Avogadro's number, A is the area of the working electrode, v is the voltage scan rate, R_G is the gas constant ($8.3144598\ \text{J}\cdot\text{mol}^{-1}\cdot\text{K}^{-1}$), T is the temperature in K, and c is the bulk analyte concentration.

7.2 Chronoamperometry and Circuit Analysis

In **chronoamperometry**, a step voltage (or a square wave) with different amplitudes ($V(t)$) is sent to an electrochemical cell which contains the solution and two electrodes. In order to simplify the discussion of this technique, a **Randles circuit** is used to model the electrochemical cell, as illustrated inside the dashed box in Fig. 7.2a. In general, R_1 represents the resistance of the solution and the contacts, and R_2 and C are the resistance and the capacitance of the electrical double layer (EDL), respectively.

As the circuit model for electrochemical systems could be very complicated for various applications, it is useful to review the techniques for circuit analysis. Common methods on the AC circuit analysis include solving a set of differential equations or performing

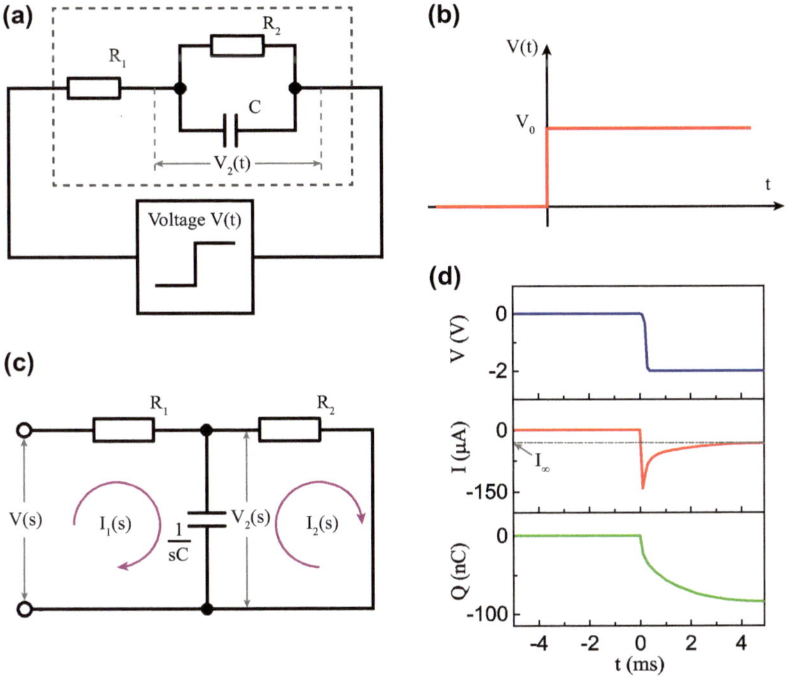

Figure 7.2 (a) The circuit diagram of a two-electrode electrochemical cell for chronoamperometry. The dashed box encloses a Randles circuit where a resistor (R_1) is in series with a component which consists of a second resistor (R_2) and a capacitor (C) in parallel. $V(t)$ is a step voltage source, and $V_2(t)$ represents the voltage across C or R_2. (b) The voltage $V(t)$ shows a step function in time domain. (c) Equivalent circuit of (a) in the complex frequency (s) domain for mesh analysis. (d) Measured total voltage (V, blue) and current (I, red) of the circuit, and the calculated stored charges (Q, green) on EDL.

Laplace transforms, on the basis of Kirchhoff's laws [Alexander and Sadiku (2012)]. Both techniques give the same result. When the circuit is not too complicated, there is not much difference in using either one of the methods, in terms of the workload of calculation. However, the number of differential equations increases rapidly with the complexity of the circuit. This significantly complicates the calculation. It is therefore more practical to analyze the circuit using Laplace transform.

Laplace transform facilitates the calculations in many scientific and engineering disciplines, as it is able to turn differential equations (in time domain) into algebraic equations (in frequency domain). The procedure for circuit analysis using Laplace transform generally includes four steps. (1) Find the Laplace transform of the input signal. (2) Find the circuit element model in frequency (s) domain (i.e., find the impedance of each element in s-domain). (3) Apply Kirchhoff's laws in s-domain and find the expression of the desired quantity. (4) Find the inverse Laplace transform of the obtained expression.

As shown in Fig. 7.2b, a step function is used here to describe the input voltage ($V(t)$) for the simplicity of calculation, instead of a square wave function. It should be noted that these two waveforms have the same impact on the circuit when the square wave pulse is long enough. Therefore, the results of the circuit analysis are not affected by the selection of a step function for the voltage source. Figure 7.2b shows the step function of the input voltage in time domain, where $V(t) = V_0$ when $t > 0$. The equivalent circuit diagram of Fig. 7.2a for the investigation in frequency domain using mesh analysis is shown in (c), where $I_1(s)$ and $I_2(s)$ are the currents looping in the same direction for mesh 1 and 2, respectively [Alexander and Sadiku (2012)].

It has been shown that the Laplace transform for the input voltage, the resistors, and the capacitor are $V(s) = \frac{V_0}{s}$, $R_1(s) = R_1$, $R_2(s) = R_2$, and $C(s) = \frac{1}{sC}$, respectively [Alexander and Sadiku (2012)]. Hence the mesh analysis based on Kirchhoff's laws gives

$$V(s) = \frac{V_0}{s} = I_1\left(R_1 + \frac{1}{sC}\right) - I_2\left(\frac{1}{sC}\right), \tag{7.2a}$$

$$0 = -I_1\left(\frac{1}{sC}\right) + I_2\left(R_2 + \frac{1}{sC}\right), \tag{7.2b}$$

where $I_1 = I_1(s)$ and $I_2 = I_2(s)$. Rearranging above equations and multiplying by R_2, we obtain

$$V_2(s) = I_2R_2 = \frac{V_0R_2}{s[(R_1R_2C)s + (R_1 + R_2)]} = \frac{V_0\frac{1}{R_1C}}{s\left(s + \frac{1}{R_xC}\right)}, \tag{7.3}$$

where $R_x = \frac{R_1R_2}{R_1+R_2}$. The inverse Laplace transform of this equation gives the voltage across the capacitor in time domain ($V_2(t)$).

In general, the inverse Laplace transform of a function in s-domain ($F(s)$) can be directly obtained by an integration $f(t) = \mathcal{L}^{-1}[F(s)] = \frac{1}{2\pi i}\int_{\sigma-i\infty}^{\sigma+i\infty} F(s)e^{st}\mathrm{d}s$, where i is the imaginary number and σ is a real number which allows the integration to be in the region of convergence of $F(s)$.

The workload of the calculation will be quickly escalated with the complexity of $F(s)$. Therefore, it is convenient to break $F(s)$ into simple terms whose inverse Laplace transform can be found in mathematical handbooks, and then find the inverse Laplace transform for $F(s)$. *Heaviside's theorem* is a well-known technique which involves two steps. (1) Decompose $F(s)$ into simple terms by partial fraction expansion. (2) Compose the inverse Laplace transform of each term by looking up a mathematical handbook.

For example, if $F(s)$ is in the form of

$$F(s) = \frac{N(s)}{(s+p_1)(s+p_2)\cdots(s+p_n)}, \tag{7.4}$$

where $N(s)$ is a function of s. By partial faction expansion, it can be re-written as

$$F(s) = \frac{k_1}{s+p_1} + \frac{k_2}{s+p_2} + \cdots + \frac{k_n}{s+p_n}, \tag{7.5}$$

where $k_1, k_2, \cdots, k_n$ are the residues of $F(s)$, which can be found in many different ways. By the residue method, $k_i = (s+p_i)F(s)|_{s=-p_i}$. Note that k_i is not necessarily 0 although $s + p_i = 0$, as $F(s)$ could be a very large number (for example, $\frac{1}{s+p_i}$). In this case, the inverse Laplace transform for $F(s)$ is

$$f(t) = (k_1e^{-p_1t} + k_2e^{-p_2t} + \cdots + k_ne^{-p_nt})u(t), \tag{7.6}$$

where $u(t)$ is the unit step function which is the same as $\frac{V(t)}{V_0}$.

Following these steps, Eq. 7.3 can be re-written as

$$V_2(s) = \frac{k_1}{s} + \frac{k_2}{s+\frac{1}{R_xC}}, \tag{7.7}$$

where

$$k_1 = sV_2(s)|_{s=0} = \frac{V_0\frac{1}{R_1C}}{\frac{1}{R_xC}} = V_0\frac{R_x}{R_1},$$

and

$$k_2 = \left(s+\frac{1}{R_xC}\right)V_2(s)\Big|_{s=-\frac{1}{R_xC}} = \frac{V_0\frac{1}{R_1C}}{-\frac{1}{R_xC}} = -V_0\frac{R_x}{R_1}.$$

The voltage across the capacitor in time domain is then obtained

$$V_2(t) = V_0\frac{R_x}{R_1}\left(1 - e^{-\frac{t}{R_xC}}\right) = V_{2\_\infty}\left(1 - e^{-\frac{t}{R_xC}}\right), \tag{7.8}$$

where $V_{2\_\infty} = V_2(t \to \infty) = V_0\frac{R_x}{R_1}$.

Hence, the total current in the circuit can be written as

$$i(t) = \frac{V_0}{R_1} - \left[V_0\frac{R_x}{R_1^2}\left(1 - e^{-\frac{t}{R_xC}}\right)\right]. \tag{7.9}$$

It can be seen that the total current becomes stable when $t \to \infty$, i.e., $i_\infty = i(t \to \infty) = \frac{V_0}{R_1+R_2}$. Subtracting this term from Eq. 7.9 and then performing the integration of it over time, we obtain

$$\begin{aligned} Q_x &= \int [i(t) - i_\infty]\,\mathrm{d}t \\ &= \int \left[\frac{V_0}{R_1}\left(1 - \frac{R_x}{R_1} - \frac{R_x}{R_2}\right) + \frac{R_xV_0}{R_1^2}e^{-\frac{t}{R_xC}}\right]\mathrm{d}t \\ &= \int \frac{R_xV_0}{R_1^2}e^{-\frac{t}{R_xC}}\,\mathrm{d}t \\ &= \frac{R_x}{R_1}CV_{2\_\infty} \\ &= \frac{R_x}{R_1}Q_{2\_\infty}, \end{aligned} \tag{7.10}$$

where Q_{2_∞} is the charge stored on the capacitor when $t \to \infty$. It can be seen that the result of this integration (Q_x) represents the charges on the capacitor, only by the difference of a factor of $\frac{R_x}{R_1}$.

Figure 7.2d shows the measured total current (I, red) of a two-electrode electrochemical cell, where a step voltage (V, blue) is applied. Q (green curve, same as Q_x in Eq. 7.10) in the figure is obtained by integrating the current over time, i.e. $Q = \int(I - I_\infty)\mathrm{d}t$, where I_∞ is the total current in the circuit when $t \to \infty$. This can be used to estimate the electrical properties of EDL, including the stored charges (Q_{2_∞}) and the differential capacitance (obtained by finding the derivation $\frac{\mathrm{d}Q_{2\_\infty}}{\mathrm{d}V}$).

It should be noted that the redox current has not been taken into account in this section, which could appear in the measured current in addition to the charging behavior. For the cases where the redox current is significant, a modification to the analysis is necessary. Readers are encouraged to consult [Bard and Faulkner (2000)] for more details.

7.3 Electrochemical Impedance Spectroscopy

The experimental techniques described in the previous sections impose large potentials at the electrode–electrolyte interfaces, and hence drive the studied electrochemical systems to a condition away from the equilibrium. The electrochemical impedance spectroscopy (EIS) investigates the non-Faradaic processes by sending a small-amplitude sinusoidal voltage wave across the interface, which is able to keep the perturbation to the system at a minimum level. In addition, the method allows to study the frequency-dependent behavior of the solution by varying the frequency in a broad range. The obtained response can be treated theoretically by linearized current-potential characteristics. These advantages makes EIS a very popular experimental technique for the study of non-Faradaic processes in electrochemistry.

7.3.1 Electric Circuit Elements

Before the discussion of circuit modeling of EIS for the processes at electrode–electrolyte interfaces, it is useful to review the conventional electric circuit elements. There are three essential linear circuit elements in the discussion for classical electrical circuits, i.e., resistor (R), capacitor (C), and inductor (L). Note that R, C, and L are also used to define the resistance, capacitance, and inductance of these circuit elements. They can be described using the differential equations

$$V = IR, \tag{7.11a}$$

$$I = C\frac{\mathrm{d}V}{\mathrm{d}t}, \tag{7.11b}$$

$$V = L\frac{\mathrm{d}I}{\mathrm{d}t}, \tag{7.11c}$$

where V and I are the voltage and the current across the circuit elements.

In an AC circuit which consists of many electrical components, the differential equations could be very complicated and they might be difficult to solve. It is then useful to define the frequency dependent impedance (Z) or admittance (Y) for the circuit elements, where

$$V = IZ = \frac{I}{Y}. \tag{7.12}$$

A sinusoidal alternating voltage can usually be expressed by

$$V(t) = V_0 e^{j\omega t}, \tag{7.13}$$

where V_0 is the voltage amplitude, ω is the frequency, and $j = \sqrt{-1}$ is the imaginary unit. Substitution of Eq. 7.13 into Eq. 7.11 gives the expression of I, which can be used to derive Z and Y. As j is present in V and I, Z and Y could be complex numbers, which can be expressed as

$$Z = Z' + jZ'', \tag{7.14a}$$

$$Y = Y' + jY''. \tag{7.14b}$$

Z' and Y' are the real part, Z'' and Y'' are the imaginary part for Z and Y. In EIS, the phase of Z is usually defined as

$$\phi = \arctan \frac{-Z''}{Z'}. \tag{7.15}$$

The phase not only describes the ratio between the imaginary and real parts of the impedance, but also indicates the phase difference between the obtained current and the applied voltage (due to the definition of Z, Eq. 7.12).

The impedance and admittance for a resistor can be derived,

$$Z_R = R, \tag{7.16a}$$

$$Y_R = \frac{1}{R}. \tag{7.16b}$$

Note that Z_R and Y_R are real numbers for resistors. Hence, the phase in this case is 0.

The impedance and admittance for a capacitor are written as

$$Z_C = -\frac{j}{\omega C}, \tag{7.17a}$$

$$Y_C = j\omega C. \tag{7.17b}$$

It can be seen that both Z_C and $_CY$ are pure imaginary numbers. The phase is then $\frac{\pi}{2}$, as $-Z''_C > 0$.

In the case of inductor, the impedance and admittance are

$$Z_L = j\omega L, \tag{7.18a}$$

$$Y_L = -\frac{j}{\omega L}. \tag{7.18b}$$

The calculation shows that $\phi = -\frac{\pi}{2}$ for L, since $-Z_L'' < 0$ here.

Note that although the impedance and admittance are obtained by the simple algebraic equation (Eq. 7.12), a more rigorous derivation using Laplace transform discussed in the previous section gives the same result.

7.3.2 Specific Circuit Elements

The previously discussed elements from classical electric circuits can be adopted to model most simple electrochemical processes. This is because ions in electrolyte are treated as simple charge carriers like the electrons in solid state materials, where their transport can be described by the classical electric conduction models. However, due to the finite size and the greater mass of the ions, the transport of ions in liquid would be affected by other factors such as diffusion, ionic structure at electrode–electrolyte interfaces (electrical double layer), ion concentration, temperature, stress, viscosity, among others. The conventional electrical circuit elements are not able to describe these processes/properties, specific circuit elements for electrochemistry are, therefore, required for the investigation of the electrochemical processes using EIS.

A well-known circuit element used to describe the electrical double layer capacitance is the constant phase element (Q), instead of C. This is because the frequency-dependent behavior of the electrical double layer capacitance shows a dispersion effect from the classical C. The physics mechanism of this, however, is not clear. This does not stop people from using Q to model the EIS result. It is commonly accepted that the impedance and admittance of the constant phase element are defined as

$$Z_Q = \frac{1}{Y_0(j\omega)^\alpha} = \frac{1}{Y_0\omega^\alpha}e^{-j\frac{\alpha\pi}{2}}, \tag{7.19a}$$

$$Y_Q = Y_0(j\omega)^\alpha = Y_0\omega^\alpha e^{j\frac{\alpha\pi}{2}}, \tag{7.19b}$$

where Y_0 and α are two arbitrary parameters used in circuit modeling for the analysis of EIS. The unit for Y_0 is $\Omega^{-1}\cdot\text{cm}^{-2}\cdot\text{s}^{-\alpha}$, while α is a dimensionless number. Q is called constant phase element, because the phase ϕ,

$$\phi = \frac{\alpha\pi}{2}, \tag{7.20}$$

is a constant, due to its definition (Eq. 7.15). Note that this phase is the same as that for a conventional capacitor if $\alpha = 1$.

If the mass transfer process needs to be considered during an EIS measurement, for example, if a Faradaic current occurs, the **Warburg impedance** should be used. It accounts for the semi-infinite diffusion process of the molecules/ions in the stagnant layer of the solution. The Warburg impedance for an electrochemically reversible process is then defined as

$$Z_W = \frac{\gamma R_G T}{n^2 F^2 c\sqrt{2\omega D}}(1 - j), \tag{7.21}$$

where γ is the reaction order, R_G is the gas constant, T is the temperature in K, n is the number of electrons involved in a single reaction, F is the Faraday constant, c is the bulk analyte concentration, and D is the diffusion coefficient. It is interesting to note that the real and the imaginary parts have the same magnitude in Eq. 7.21, which indicates that *the phase for Z_W is $\frac{\pi}{4}$*. This is a very useful information when analyzing the EIS results, as a straight line in the Nyquist chart (defined in section 7.3.3) can be observed. Note that the Warburg impedance can also be used to describe the irreversible process, except that Eq. 7.21 needs to be re-defined.

In addition to the semi-finite diffusion discussed here, various circuit elements are developed for modeling different diffusion processes, which include finite diffusion, spherical diffusion, cylindrical diffusion, and so on. Readers who are interested in these topics are recommended to read [Lasia (2014)].

7.3.3 Equivalent Circuits

These circuit elements discussed in sections 7.3.1 and 7.3.2 are particularly useful for the analysis of the electrochemical processes using equivalent circuits. When measuring the impedance of the electrochemical cell using EIS, the obtained current is used to calculate the complex impedance of the system. This can be studied by comparing the experimental results with the theoretical prediction of different electrical circuits. The equivalent circuit which shows the similar frequency-dependent impedance behavior could be used to model the system if each of the circuit elements represent a meaningful physical process, for example, the charge transfer or the electrical double layer capacitance. Therefore, EIS is a powerful tool for the investigation of the electrical properties, and it can be adopted to provide the interpretation of the physical processes. Here we focus on the discussion of three fundamental circuits which only involve a resistor and a capacitor. This is because these widely accepted circuit elements are directly relevant to probing the electronic properties in electrode. It is worth noting that, based on the working principles shown in this section, readers should be able to extend the discussion of circuits to more complicated ones which include the Warburg and the constant phase elements presented in section 7.3.2.

7.3.3.1 Impedance plots

It is generally impossible to use only one circuit element to describe a complicated electrode–electrolyte system. The equivalent circuit may comprises of many different circuit elements. Hence, the total impedance/admittance of the whole circuit is complicated. It is then useful to plot the frequency-dependent impedance/admittance in graphs for analysis. Two popular plots that people often use to study the impedance/admittance are the Nyquist plot and the Bode chart. The **Nyquist plot** shows the real (as x axis) and imaginary (as y axis) parts of the impedance/admittance in a same graph. The **Bode chart** plots the magnitude and the phase of the impedance/admittance as a function of frequency.

Figure 7.3 shows the Nyquist plot and the Bode chart of a capacitor ($C = 1$ F). Note that $-Z''$ is used as the y axis for the Nyquist plot, instead of Z''. The logarithmic scale is adopted to show the relationship between the magnitude of the impedance ($|Z_C|$) and frequency (f). As described by Eq. 7.17, the real part of Z_C is 0 and the phase is $\frac{\pi}{2}$ (90°), which is shown schematically in Fig. 7.3. This indicates that the absolute value of the impedance of a capacitor

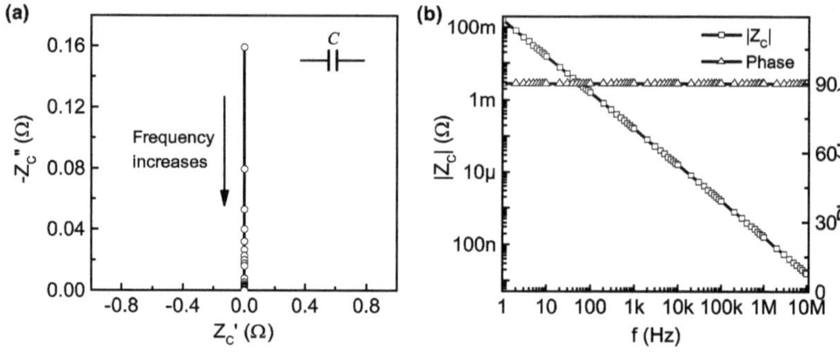

Figure 7.3 (a) The Nyquist plot of a capacitor ($C = 1$ F), where the real (Z') and imaginary ($-Z''$) parts of the impedance are shown. Inset: the symbol of a capacitor in circuits. (b) The Bode chart of the same capacitor, where the magnitude ($|Z_C|$) and the frequency (f) are shown in logarithmic scale, and the phase is in linear scale.

decreases at higher frequencies, this is because the displacement current of the capacitor ($I = C\frac{\mathrm{d}V}{\mathrm{d}t}$) increases with the frequency. The phase difference between the obtained current and the applied voltage is fixed (90°). Therefore, these plots provide an intuitive tool to study the impedance of electrochemical systems.

7.3.3.2 Electrical circuits

It is then useful to look at the impedance plots for some of the basic electric circuits. The examples given in this section include resistor-capacitor (RC) series and parallel circuits, as these are very common in most research projects. In a series circuit, the total impedance is the sum of the impedance of each circuit element, while in parallel circuit the total admittance is the sum of the admittance of each circuit element.

For a RC series circuit, the total impedance is then

$$Z = R - \frac{j}{\omega C}. \tag{7.22}$$

Hence, the magnitude of Z and the phase (ϕ) are

$$|Z| = \frac{\sqrt{1 + (\omega RC)^2}}{\omega C}, \tag{7.23a}$$

$$\phi = \arctan \frac{1}{\omega RC}. \tag{7.23b}$$

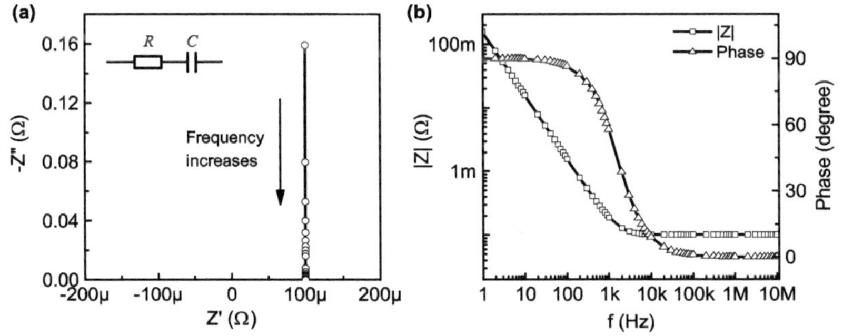

Figure 7.4 (a) The Nyquist plot and (b) the Bode chart of a *RC* series circuit ($C = 1$ F, $R = 100$ μΩ). Inset in (a): *RC* series circuit.

Figure 7.4 shows the Nyquist plot and the Bode chart of a *RC* series circuit ($C = 1$ F, $R = 100$ μΩ), calculated using Eqs. 7.22 and 7.23. It can be seen from Fig. 7.4a that R shifts the impedance in the complex plane (the Nyquist plot) by the value of the resistance. This plot, however, does not give the frequency-dependent behavior.

The magnitude of the impedance and the phase in Fig. 7.4b show a different response at higher frequencies, in comparison with Fig. 7.3b. The impedance stabilizes at a value determined by R, and the phase changes from 90° to 0° (which is the phase of R). This is because a capacitor behaves as a good conductor (with very low impedance) at high frequencies, the current flow in the circuit will then be limited by the resistance. Although the capacitor would induce a 90° phase difference between the current and the voltage, the higher impedance of R brings the phase back to 0°. This indicates that C dominates the impedance of *RC* series circuits at low frequencies, while R is dominant at high frequencies.

If R is connected to C in a parallel circuit, the total impedance can be derived by finding the inverse of the total admittance, i.e.,

$$Z = \frac{1}{Y} = \frac{1}{\frac{1}{R} + j\omega C}. \tag{7.24}$$

Re-arranging this equation gives the real and the imaginary parts of Z,

$$Z' = \frac{R}{1 + (\omega RC)^2}, \tag{7.25a}$$

$$Z'' = -\frac{\omega R^2 C}{1 + (\omega RC)^2}. \tag{7.25b}$$

The magnitude and the phase of Z can be derived as

$$|Z| = \frac{R}{\sqrt{1 + (\omega RC)^2}}, \tag{7.26a}$$

$$\phi = \arctan \omega RC. \tag{7.26b}$$

Equation 7.25 indicates that Z' and Z'' construct a semi-circle in the complex plane (the Nyquist plot) as $\left(Z' - \frac{R}{2}\right)^2 + (Z'')^2 = \left(\frac{R}{2}\right)^2$, where the center is at $\left(Z' = \frac{R}{2}, Z'' = 0\right)$ and the radius is $\frac{R}{2}$. Figure 7.5a illustrates the Nyquist plot of a RC parallel circuit, where $C = 1$ F and $R = 100$ μΩ. The impedance shows a semi-circle shape in the complex plane as discussed.

The Bode chart shown in Fig. 7.5b suggests a very different frequency-dependent behavior of the impedance for the RC parallel circuit than that for the series circuit. At low frequencies, most of the current flows through the resistor, as the capacitor can be considered as an open-circuit element (with large impedance value). Hence, the total impedance is determined by R, and the phase difference between the current and the voltage is 0°. When the frequency increases, the capacitor behaves like a good conductor with low impedance. Therefore, the impedance due to capacitance dominates in the parallel circuit at high frequencies, i.e., $|Z|$ decreases and the phase increases to 90°.

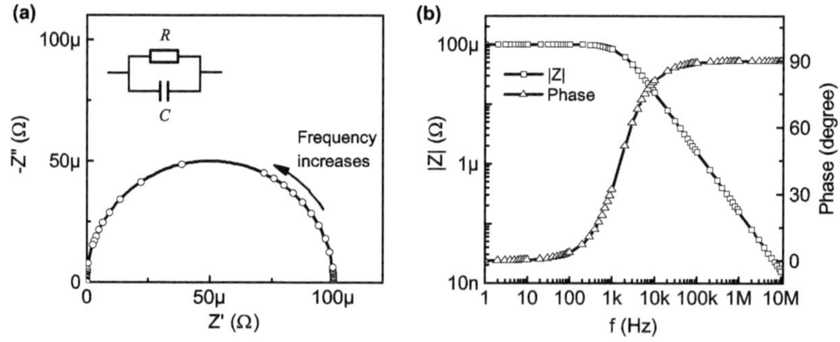

Figure 7.5 (a) The Nyquist plot and (b) the Bode chart of a RC parallel circuit ($C = 1$ F, $R = 100$ μΩ). Inset in (a): RC parallel circuit.

In experiments, the circuit model is generally more complicated. However, the methods of analysis is still similar to the discussion in this section, except that more circuit elements may need to be considered, for example, the constant phase element (Q), Warburg impedance (W), and so on. The discussion for the circuit models including these circuit elements are not discussed here. Readers who are interested are encouraged to take it as a practice.

It should also be noted that, on some occasions, the obtained impedance of an electrode–electrolyte system could be represented using various circuit models, which may be drastically different from each other. It is then necessary to find a physically meaningful circuit for the studied system. The Randles circuit shown in Fig. 7.2 (enclosed by the dashed box) is a commonly used circuit model for EIS, where the capacitor (C) may be replaced by the constant phase element in some cases. This is because Q might be more suitable for the description of the double layer capacitance. R_1 and R_2 represent the resistances of the solution and the one across the electrical double layer, respectively.

There are many other electrochemical techniques which can be used to study electrochemical systems, for example, pulse voltammetry, hydrodynamic techniques, microelectrodes, and so on. This is a well-established field which requires a systematic study. Readers who are interested in electrochemistry are recommended to consult [Bard and Faulkner (2000)].

Chapter 8

Applications and Outlook

Many applications in various fields, including biology, electronics, and energy, involve the use of electrode–electrolyte systems. The mixed electron-ion conductive system provides a unique tool to investigate the electron-ion interaction (such as charge transfer and mobility alteration). The fast and sensitive electrons can be utilized to probe the ionic and electrical activities in the solution, while the free ions in high dielectric constant (κ) liquid are ideal for energy storage and/or the enhancement of electronic properties in the electrode. As the topic of this book is focused on the discussion of electronic properties at graphene–electrolyte interfaces, applications in energy storage based on the ionic properties are not discussed here.

As electrons move faster in graphene than most of the known materials, due to the very high mobility, graphene is among the perfect candidates as the electronic conductor for the probe of ionic activities at graphene–electrolyte interfaces. As discussed in Chapter 3, although graphene can be considered as a semiconductor (because the Fermi level sits near the "Dirac" point in an undoped graphene), the absence of the band gap renders graphene as a very sensitive platform for the ions in the environment. Graphene–electrolyte systems can then be used to develop the applications for

Graphene–Electrolyte Interfaces: Electronic Properties and Applications
Hualin Zhan

ISBN 978-981-4774-94-9 (Hardcover), 978-1-003-04487-1 (eBook)
www.jennystanford.com

biosensing or neural activity recording. Note that the liquid, on the other hand, can be used to improve the performance of the graphene electronic devices.

8.1 Conventional Electrical Biosensing

Detection of molecules in liquid at low concentration has been of great interest for life science applications including medicine, ecology and biology. Developing sensitive, fast and label-free biosensors in solution is one of the main goals in these fields [Kostarelos and Novoselov (2014b)]. There are various types of biosensors developed based on different sensing mechanisms, including optical (fluorescence or plasmonic), mechanical (cantilever-based), and electrical (amperometric and potentiometric). This chapter focuses on the discussion of the electrical biosensors using graphene. The **amperometric** biosensors detect ions/molecules by measuring the current across the graphene–electrolyte interface when a voltage is applied. Cyclic voltammetry is a common amperometric technique. The **potentiometric** method, on the other hand, measures the ion- or molecule-induced potential change to identify the chemical species. Ion-sensitive field-effect transistors are developed based on this principle.

8.1.1 Cyclic Voltammetry

As described in section 7.1, cyclic voltammetry (CV) can be used to detect the chemical species in a solution by identifying the redox peaks and measuring current in the obtained voltammogram. Molecular detection by CV is a classical technique which has been studied intensively. Various methods are developed to interpret the voltammogram for analysis of the reaction type and molecular species. The method introduced in section 7.1 is an example. Note that CV is mainly used to detect electrochemically active chemicals. For the chemical species which are not involved in electrochemical reactions, it is impossible to detect directly. Modification to the electrode and/or introduction of catalysts and mediators are then required for the detection of electrochemically inactive species (such

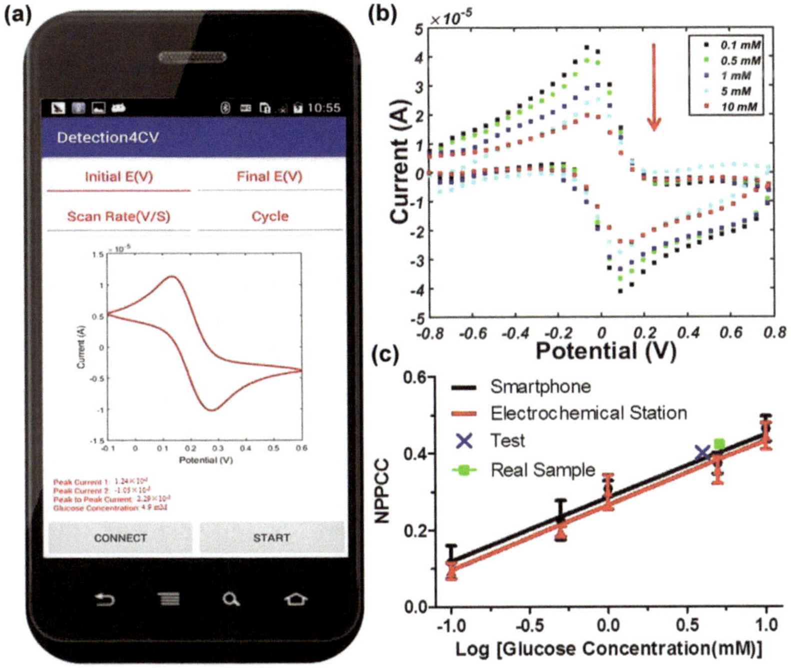

Figure 8.1 (a) Photograph of a smartphone based cyclic voltammetry system, where a CV curve is shown. (b) Detection of glucose at different concentration by CV, where graphene (reduced graphene oxide) modified screen printed electrodes are used. (c) The normalized peak-to-peak current change (NPPCC) of the current in the cyclic voltammogram, obtained by the smartphone (black line) and a commercially available electrochemical station (red line). The blue cross and the green square show the results tested in a lab-prepared glucose solution and a blood sample. Reprinted with permission from [Ji et al. (2017)] Copyright Elsevier.

as glucose). The commonly used catalyst and the mediator for the amperometric detection of glucose are glucose oxidase (GOx) and ferrocenemethanol.

More interestingly, recent development shows that the electrochemical potentiostat can be integrated in smartphones, which delivers a fast, convenient, and low-cost tool to the patients for monitoring the concentration of certain molecules in their bodies. Figure 8.1a shows an example of a smartphone based cyclic voltammetry system for biosensing, where graphene (reduced graphene oxide) modified screen printed electrodes are used [Ji et al. (2017)]. The voltammogram shown in the screen of the phone is a real-time measurement of CV. The detection of glucose at different concentration in the mM range is demonstrated in Fig. 8.1b. It can be seen that the magnitude of the current decreases as the concentration of glucose increases, which proves that this technique is effective for the detection of glucose.

Figure 8.1c summarizes the normalized peak-to-peak current change (NPPCC) of the current in the cyclic voltammogram, when the liquid containing glucose of different concentration is tested. NPPCC is defined as $\frac{I_b - I_a}{I_b}$, where I_a and I_b are the peak-to-peak current change for the solution with and without the glucose. It can be seen that the smartphone based system shows a comparable performance in comparison with a commercially available potentiostat. This system is used to test the glucose in a lab-prepared solution (blue cross) and a blood sample (green square), indicating the potential of commercializing this technique.

8.1.2 Ion-Sensitive Field-Effect Transistors

The amperometric biosensing technique discussed previously uses graphene as a conductor. This, however, has not fully explored the sensitive nature of the graphene to the environment. The electrons in a monolayer (or few-layered) graphene respond rapidly to the environment (such as the ions/molecules in the liquid) due to the linear gapless band structure. This response could be reflected in measurable electrical properties such as the sheet conductance. The sensitivity of this response is generally relevant to the quality of graphene and the number of monolayers. It is then possible to improve the sensitivity of the biosensing techniques using monolayer (or few-layered) graphene in the graphene–electrolyte systems.

Field-effect transistor is a device where the electrical current between two contacts (termed as source and drain) can be modulated by the electric potential applied on a gate electrode (i.e., the field effect), as discussed in section 2.3.3. The conductor between the source and drain is defined as the channel, and it is a key component in the design of a transistor, as it must respond to the external voltage (i.e., the charge carriers in the channel can be modulated by the applied electric field). As the number of charge carriers can be readily tuned when they are confined in the certain dimensions in the nanostructures, the channel of a transistor is often made of a low-dimensional material (Figs. 2.5a–c). For example, the channel for a single electron transistor is a quantum dot. Graphene, as a 2D material with the tunable Fermi level in the gapless band structure, is also ideal for the development of field-effect transistors (section 3.3.1). If the field effect is induced by the ions/molecules in the environment, the conductance change of the graphene transistor can be used as the indicator for this chemical species.

The field-effect modulation of charge carriers in graphene of a transistor can be considered as a capacitive charging/discharging process. Conventionally, the capacitance is provided by a "parallel-plate" capacitor where an oxide layer is placed between the graphene conduction channel and gate electrode as the dielectric medium (Figs. 3.9a–b). The net density of charge carrier in graphene conduction channel would increase/decrease during charging/discharging between the graphene-oxide-gate capacitor, i.e., field effect. This field effect in solid materials can be transferred to liquids, where the oxide layer is replaced by electrolyte. Since the electrical double layer with a large capacitance value (due to the thin Stern layer, section 4.3) is formed at the electrode–electrolyte interface, the charge carriers in the channel can be modulated using the electrical double layer. As the gate electrode is usually immersed in the liquid to produce the electrical double layer at the graphene–electrolyte interface when a voltage is applied, this configuration is defined as the **liquid gated** setup. A transistor with a liquid gate is then called a liquid gated transistor, or ion-sensitive field-effect transistor (ISFET). This is because the resistance of the conductive channel of the transistor is sensitive to the potential induced by ions/molecules in liquid in such a liquid gated configuration [Moss et al. (1975)]. This technique is then also regarded as a potentiometric method. Note that the liquid-gate configuration provides an ideal platform for electrical operation in aqueous solutions, as electrical double layer at the graphene–electrolyte interface reduces the gate voltage from tens of volts in all-solid graphene devices to less than a volt [Novoselov et al. (2004); Zhan et al. (2017)]. The low gate voltage in aqueous solutions ensures that the undesired redox reactions of water do not happen.

Graphene transistors have attracted significant attention in the detection of gas molecules in recent years due to the high sensitivity [Schedin et al. (2007)]. Since ISFET measures the electrical resistance change of the conducting channel of the transistor after interaction with the analyte molecules [Yan et al. (2014)], graphene transistors are adopted in liquid for the detection of ions/molecules. An advantage of ISFET in biosensing is that it does not require the redox reactions in liquid, as the biosensor detects the ions/molecules induced potential, instead of the redox current.

Figure 8.2a illustrates the design of a graphene ISFET for glucose detection [Zhang et al. (2015)]. It can be seen that a current flows from the drain (D) to source (S) in the monolayer graphene channel when a drain-source voltage (V_{DS}) is applied. The gate voltage (V_{G}) is applied between the gate electrode (G) and S, where the gate electrode, in this case, is also made of graphene. The graphene channel between D and S is decorated with glucose oxidase (GOx), platinum (Pt) nanoparticles, and polymer matrix, in order to selectively detect the electrical signal given by glucose (with hydrogen peroxide, H_2O_2).

The blue curve in Fig. 8.2c shows the measured drain-source current (I_{DS}) as a function of the gate voltage (V_{G}) of a graphene ISFET in a phosphate-buffered saline (PBS) solution. The "Dirac" point represented by the minimum of I_{DS} is located at a gate voltage around 0.3 V, indicating that the graphene channel could be p-doped in this case. When a 1 mM glucose is added into the electrolyte, the "Dirac" point shifts to the left where V_{G} is less than 0.2 V (red curve in Fig. 8.2c). This indicates that the graphene channel may become less p-doped in the presence of glucose. This change to the resistance of the graphene channel can then be used to detect glucose. Figure 8.2d demonstrates the ability of this graphene ISFET for the detection of glucose of the concentrations in the μM range. In comparison with Fig. 8.1c, the sensitivity of the ISFET developed using a

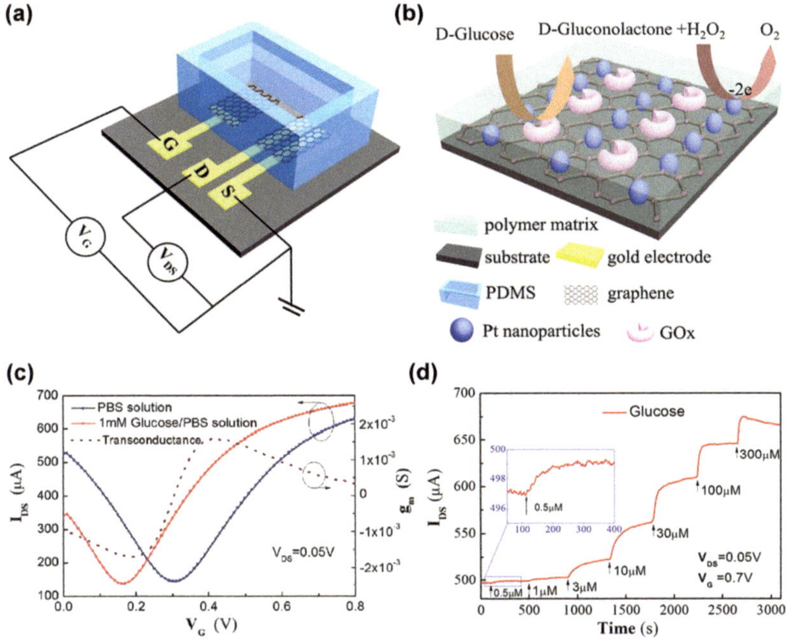

Figure 8.2 (a) Design of a graphene ion-sensitive field-effect transistor where an additional graphene is immersed in liquid as gate. (b) Illustration of the graphene channel in an ISFET for the detection of glucose, which is decorated with glucose oxidase (GOx), platinum nanoparticles, and polymer matrix. (c) The measured drain-source current (I_{DS}) as a function of the gate voltage (V_G) of a graphene ISFET in a phosphate-buffered saline (PBS) solution with (red) and without (blue) glucose. The shown transconductance (g_m) is the derivative of I_{DS} over V_G for the blue curve. (d) The change of I_{DS} with the addition of glucose in the PBS solution, where the drain-source voltage (V_{DS}) and the gate voltage are 0.05 V and 0.7 V, respectively. Reprinted with permission from [Zhang et al. (2015)] Copyright Nature Publishing Group.

monolayer graphene is higher than the amperometric method which simply uses reduced graphene oxide as an electrode. It is worth to mention that, the selectivity of graphene ISFET is done by decorating graphene with functional groups, the functionalization process is then usually required for the development of the transistor-based biosensors.

8.2 Liquid-Gated Hall-Effect Biosensing

Further improvement of the sensitivity of the ISFETs made of unfunctionalized graphene, particularly in liquid, has been a debatable topic [Fu et al. (2011); Kumar et al. (2013); Min et al. (2011)]. The resistance change of pristine graphene ISFETs in response to some low concentrated solutions may not be observable [Fu et al. (2011); Moss et al. (1975)], although a high-quality pristine graphene would in principle provide high sensitivity. In contrast to this intuition, some research suggests that pre-treatment of graphene may improve the sensitivity, regardless the defects that this process may introduce, such as depositing molecular binding sites [Dankerl et al. (2010); Fu et al. (2013); Yan et al. (2014)] or growing an extra atomic layer on graphene [Fu et al. (2011)]. However, a further thought suggests that the problem with using electrical conductance (σ) measurement for molecular sensing in an unfunctionalized or pristine ISFET may be related to the fact that it is the product of carrier charge density (en) and mobility (μ), $\sigma = en\mu$, where e is the electronic charge and n is the carrier number density. Hence although σ does not vary with the analyte concentration in some cases, n and/or μ could still change evidently. For example, if n increases by a factor of α, i.e., $n_2 = \alpha n_1$, while μ decreases by a factor of $\frac{1}{\alpha}$, i.e., $\mu_2 = \frac{\mu_1}{\alpha}$, the electrical conductance remains the same ($\sigma_2 = \sigma_1$). An ISFET in such a case is not able to detect the molecules as the electrical conductance of the channel does not change. However, the changes of the carrier density and the mobility of the channel induced by the molecules can be captured by Hall measurements.

Hall effect measurements of graphene in liquid were performed for different purposes, such as determining the carrier density of graphene [Ye et al. (2011)] and studying the charge scattering mechanism [Newaz et al. (2012)]. As these properties indicate the interaction between the graphene and the solvent molecules in solution, Hall effect measurements of graphene in liquid can be used as a method to detect molecules [Zhan et al. (2018)]. Similar to ISFETs, the gate electrode is immersed in the solution containing the analyte of interest. This liquid gated Hall measurement (LGHM) technique allows the exclusion of contact resistances and directly measure the Hall resistance (R_{H}) and the sheet resistance (ρ_{xx}) of the sensor channel (graphene) during the exposure of the sensor to analytes in liquids. The simultaneous measurement of Hall resistance R_{H} and ρ_{xx} permits the derivation of the charge carrier density ($n_{2\mathrm{D}}$) and the mobility (μ) of graphene. Therefore, in the case where no resistance change can be detected in response to the analyte concentration, Hall effect measurements may capture the variations in $n_{2\mathrm{D}}$ and μ separately. LGHM performed at various gate potentials could exhibit higher sensitivity without functionalizing graphene, in comparison with the conventional methods including CV and ISFET under the same conditions.

8.2.1 Detection of L-Histidine in the pM Range

Figure 8.3a shows a photograph of a fabricated graphene Hall structure, where D and S represent the drain and source. W and L are the width and the length of the Hall bar. $W = L = 200$ μm, as the noise is lower for larger liquid gated graphene devices [Heller et al. (2010)]. The side view of the LGHM setup is illustrated in

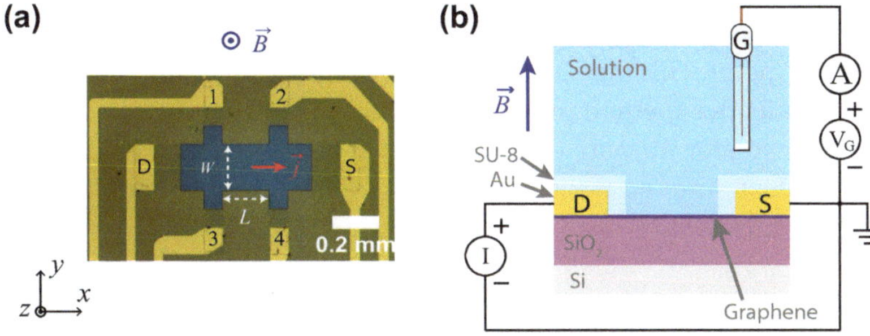

Figure 8.3 (a) A photograph of a fabricated graphene Hall bar on SiO_2/Si substrate, where the surface is covered by SU-8 except for the graphene region (blue). D and S represent drain and source. (b) Schematic diagrams of the setup of liquid gated Hall measurement, where G is the gate (Ag/AgCl electrode), I, V_{G} and A are constant current source, gate voltage source, and ammeter.

Fig. 8.3b, where the contacts 1–4 are not shown. G is the gate (Ag/AgCl) electrode, and I, V_G and A are constant current source, gate voltage source, and ammeter. The graphene Hall device and solution are contained in a tube with a cap. Changes in molecular concentration due to evaporation were discounted because the measurement time was very short. I_{leak}, as the leakage current between the gate electrode and the device, was measured by the ammeter, while the voltage V_G was being applied. Since the voltage between D and S (around 20 mV) is an order of magnitude smaller than V_G, this setup can also be considered as a two-electrode electrochemical system, where the graphene sheet and the gate are the working and the counter electrodes, respectively. In the experiments, V_G is swept back and forth from −0.6 V to 0.6 V at a rate of 4 mV/s.

Figures 8.4a–b show the measured Hall (R_H) and sheet resistances (ρ_{xx}) of graphene in solutions containing L-histidine at different concentrations in the pM range by LGHM, where PBS (10^{-2}X) is used as the supporting electrolyte. Note that the changes of interest in Fig. 8.4a occur near a position where $V_G = V_D \approx 0.35$ V. V_D is the "Dirac" point represented by the peak in Fig. 8.4b, showing the transition between the hole-dominant region ($V_G < V_D$) and the electron-dominant region ($V_G > V_D$) in graphene. R_H shown in Fig. 8.4a gradually increases with the concentration of L-histidine at the gate voltage ($V_G \approx 0.39$ V) slightly above the "Dirac" point. This indicates that the measurement of R_H can be used as an effective method to detect L-histidine of these concentrations. The reason for the gate voltage of interest being slightly above V_D is that L-histidine is a positively charged molecule. It will be attracted to the negatively biased Hall structure and interact with graphene, i.e. $V_G > V_D$.

Note the sheet resistance (ρ_{xx}) remains almost unchanged for histidine of all concentrations in pM range, as shown in Fig. 8.4b. Since the transistor-based biosensors detect molecules by the change of resistance, this indicates that the sensitivity is not as high as can be achieved by R_H measurement by LGHM. It should also be noted that there is no evident peak in the I_{leak}-V_G curve, as shown in the inset of Fig. 8.4b. This suggests that the detection of pM L-histidine in PBS by CV using an unfunctionalized graphene

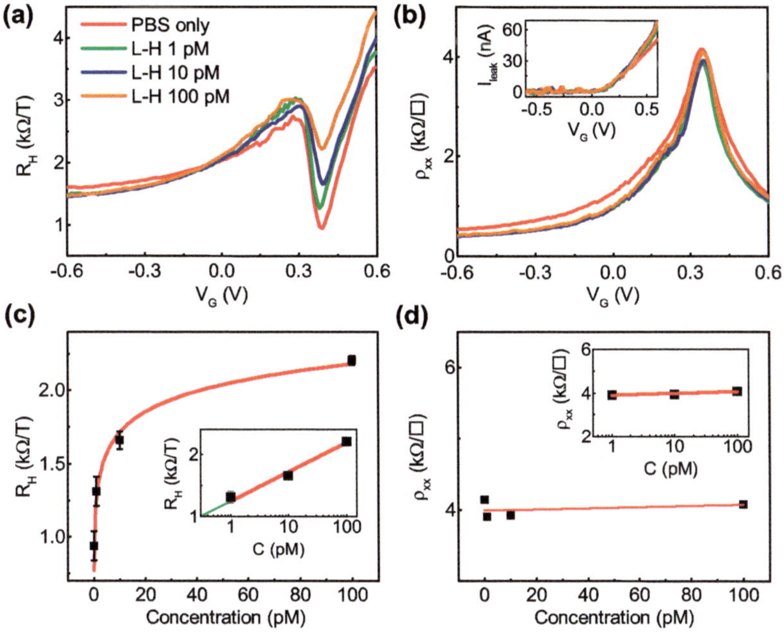

Figure 8.4 Detection of L-Histidine (L-H) by liquid gated Hall measurement of graphene at pM concentration with PBS (10^{-2}X) as the supporting electrolyte. I_{leak} is the leakage current between the gate electrode and the device. (a) Hall resistance R_{H}, and (b) sheet resistance ρ_{xx} as a function of gate voltage V_{G} (Inset: leakage current I_{leak} as a function of V_{G}). (c) R_{H}, and (d) ρ_{xx} as a function of L-histidine concentration (C) in the pM range when the gate voltage is close to the "Dirac" point (insets: plots in logarithm scale). Black squares: experimental data. Red lines: fittings (logarithm fit for (d) and linear fit for (d)). Note that the green extrapolation line in the inset of (c) indicates the sensitivity limit of this method is 300 fM.

electrode is not possible. LGHM therefore provides better sensitivity than the conventional amperometric and potentiometric methods at pM concentration of L-histidine.

Figures 8.4c–d summarize the changes in R_{H} and ρ_{xx} of graphene in pM concentration analytes near the "Dirac" point ($V_{\text{G}} \approx 0.39$ V). R_{H} follows a logarithmic dependence on concentration, while ρ_{xx} remains constant. The logarithmic relation between electronic properties of graphene with analyte concentration is also observed in other histidine detection experiments [Chen et al. (2014); Liang et al. (2011)]. The green extrapolation line in the inset of Fig. 8.4c indicates that the sensitivity limit is approximately 300 fM, where R_H is around 1 kΩ/T. This limit is determined using the maximum possible value of R_H at 0 concentration when taking into account the noise.

8.2.2 Ion-Electron Interaction

There are two interesting phenomena in Fig. 8.4a which require further discussion, as they account for the ultra-sensitivity of LGHM. (1) R_H remains the same sign in both the hole- and electron-dominant regions. This contradicts our intuition where the conventional semiconductor model suggests that R_H should have opposite signs in these regions. (2) Most of the changes of interest induced by L-histidine occur at a gate voltage near (and slightly above) the "Dirac" point where electrons and holes co-exist. The electrical properties of electrons and holes in graphene are symmetric in the absence of external perturbation. This symmetry, however, can be broken as the analyte concentration increases, which is captured by the significant change of the shape of R_H. It suggests that the ion-electron interaction should be investigated in detail to understand the obtained sensitivity of R_H.

The ions in the liquid can interact with graphene in several different ways. (i) The electrical properties of graphene are directly affected by the ionic activities in the liquid, for example, the aforementioned broken asymmetry of electron-hole mobilities. (ii) More than 30 repeating measurements performed in [Zhan et al. (2018)] confirmed that graphene becomes inhomogeneous or even get damaged when the ionic concentration or the applied gate voltage is too high. This inhomogeneity would affect the measured R_H. (iii) Ions in the solution may respond to the Hall effect measurement of graphene [Kühne et al. (2017)]. They may accumulate in the electrical contacts (1–4 in Fig. 8.3a) of the Hall structure, in the presence of magnetic field. This would offset the measured Hall field and may cause more damage to graphene as more ions stay on graphene for a longer time. The rapidly changed V_G may not give the ions enough time to rearrange themselves, hence the measured R_H contains the information from both the

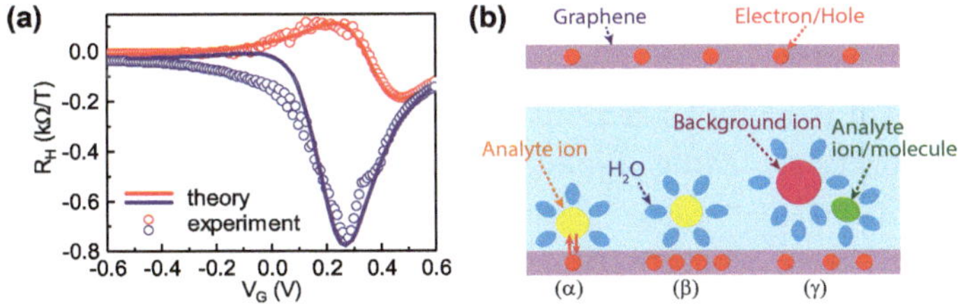

Figure 8.5 (a) Measured Hall resistance (R_H) of graphene in H_2O (red) and PBS (10^{-3}X) (violet). Empty circles: experimental results, solid lines: theoretical fittings. Note that the Fermi level (E_F) is used in the calculations, instead of gate voltage. (b) Illustration of charge carrier distribution in graphene before (upper panel) and after (bottom panel) it is in contact with liquid. Possible interactions of ions/molecules with graphene are shown, where the analyte ion/molecule is the species to be detected, and the background ion is from the supporting electrolyte.

ions in solution and electrons/holes in graphene. (iv) The quality of graphene is also important for the measurement, which can be controlled by selecting graphene with desired number of layers, grain sizes, and proper substrates. During the fabrication of the device, it is also important to minimize the time of graphene in the developing solvent (an example of the fabrication process can be found in appendix F). These steps are not only important for understanding the interaction between graphene and solution, but also critical for the repeatability and reliability of graphene devices.

As the discussion of the ionic properties in solution and the quality of graphene are outside the scope of this book, this section focus on the investigation of the change of electrical properties of graphene induced by the liquid. To explore the possible mechanism of the effect of the ions/molecules in aqueous solution on the electrons/holes in graphene, R_H of graphene is measured in DI water and in a solution containing PBS (10^{-3} X) separately, as indicated by the empty circles in Fig. 8.5a. R_H has a sinusoidal-like shape in DI water (red circles in Fig. 8.5a), which is also observed in Hall effect measurements for graphene solid-state devices without liquid [Novoselov et al. (2004)]. A reason for this is that DI water is a clean solution with only a few residual ions. Neutrally charged water molecules can be considered as a simple dielectric medium between the gate and graphene, without introducing differences between the electron and the hole properties. In the case of PBS, the shape of R_H changes drastically, as demonstrated by the violet circles in Fig. 8.5a.

Theoretical fitting (solid lines in Fig. 8.5a, calculated by Eq. 6.16b) to the measurement show that the electron-hole mobility becomes more asymmetric in PBS ($\mu_\mathrm{e} = -3.5\ \mathrm{m}^2/(\mathrm{V{\cdot}s})$ and $\mu_\mathrm{h} = 0.25\ \mathrm{m}^2/(\mathrm{V{\cdot}s})$) than that in DI water ($\mu_\mathrm{e} = -0.6\ \mathrm{m}^2/(\mathrm{V{\cdot}s})$ and $\mu_\mathrm{h} = 0.51\ \mathrm{m}^2/(\mathrm{V{\cdot}s})$). This is because ions are introduced into water in PBS. Figure 6.11c shows the calculated R_H using different values of electron-hole mobility ratio ($\mu_\mathrm{e}/\mu_\mathrm{h}$). The shape of R_H can be very different depending on this ratio, which reflects the interaction between ions and charge carriers in graphene.

Ions can interact with the electrons and holes in graphene differently in various ways depending on the charges, dipole moments, and energy levels of the ions. As shown in Fig. 8.5b, the charge carriers in graphene were distributed uniformly in the absence of external perturbation (upper panel). After the graphene is in contact with a solution, the charge carriers are re-distributed depending on the interactions between the ions and graphene. There are several possible interaction processes. (i) Direct charge transfer between the ions and graphene determined by energy level alignment (α in Fig. 8.5b), as discussed in section 5.1.1. (ii) Change of carrier mobility in graphene directly by the potential of the ions (β in Fig. 8.5b, see section 3.2.3). This is because ions, such as L-histidine, can act as charged impurities in the environment, which induce local potential distribution in graphene, produce electron-hole puddles, and change the electron/hole properties. (iii) Analyte ions/molecules with a different value of dipole moment, such as urea, can change the effect of background ions (from the environment, e.g. the supporting electrolyte) on graphene (γ in Fig. 8.5b, see the discussion of physisorption in section 5.1.1) [Newaz et al. (2012)]. This changes the electrical properties in graphene indirectly. (iv) Dielectric screening (section 5.1.2), capacitive coupling (section 5.3), and other interactions. Since these processes are not necessarily the same for electrons and holes [Lherbier et al. (2008)] and they can affect the measurement of R_H, it is suggested that LGHM can improve the sensitivity of transistor-based bio-sensors by detecting the difference of the electrical properties between electrons and holes.

8.2.3 Advantages and Disadvantages

LGHM can be used to improve the sensitivity of conventional electrical biosensing methods. The obtained R_H shows an ultra-sensitivity for L-histidine in pM concentration, while there are no observable responses by amperometric and potentiometric techniques. This is because the electrical properties, such as n_{2D} and μ, can be extracted by LGHM, which have shown clear changes in response to the change of the solution concentration. These properties however, are not captured by simple transistor-based sensors. The study in the previous subsection shows that the asymmetric electron-hole mobility induced by the charged impurities in solution plays an important role in the sensing mechanism.

Despite many promises of liquid Hall effect measurements, there are several limitations of this method which should be considered in future development. As shown in Fig. 8.3b, the setup of this method is more complicated than either CV or ISFET. However, it is believed a smaller device can be developed due to the presence of portable Hall measurement equipment. In addition, the van der Pauw method could be adopted so that the fabrication process can be facilitated (section 6.3.2.2). Similar to CV, a disadvantage is that some molecules may stay on the surface of graphene after detection (fouling). This could be resolved by some sensor surface recovery schemes such as graphene annealing, CV cleaning, and so on. Another problem is that the specificity of this technique has not been fully investigated. Although graphene is very sensitive to the environment which should interact various with analytes differently, and this could serve as the mechanism to define the selectivity. However, due to the complexity of this interaction, further studies are required (e.g., first-principal calculations, see section 5.1.1). It is then useful to decorate graphene with binding sites to improve the selectivity in some cases. In spite of these limitations, the strikingly high sensitivity of LGHM of graphene provides a promising future for utilizing this technique in application of molecular sensing.

8.3 Neural Activity Recording

Since a graphene transistor can be used to detect the potential change in electrolyte (mainly in the electrical double layer) induced by ionic activities for the application of biosensing, it can also be applied to monitor the action potentials given by the neurons in liquid. This is because the action potential fired by the neural cell introduces an extra change to the gate voltage (or the cell could be simply considered as the gate), and this will modify the conductance of the graphene channel of the transistor. The recorded conductance change can therefore be used as an indication of the neural activity.

In general, the conductance of graphene changes significantly within a small range of gate voltages for a graphene field-effect-transistor in solution (Figs. 8.2c, 8.4b, and 8.6b). This range is much smaller than that of a graphene transistor made with a solid-state dielectric layer between the gate and graphene (usually tens of volts). This indicates that even for a neural cell in a solution where the signal is in the range of mV, the change of the conductance is detectable. Figure 8.6a shows the schematics of a graphene (Gra) and a nanowire (NW) transistor for recording the activity of an embryonic chicken cardiomyocyte which is placed on top of the transistors [Cohen-Karni et al. (2010)].

Figure 8.6b shows the measured conductance of graphene as a function of the gate voltage (V_{wg}) in water, where the "Dirac" point is at 0.1 V. As the graphene transistor monitor the neural activity by detecting the conductance change, the absolute value of the slope of the red curve in Fig. 8.6b determines the sensitivity of the device. Therefore, the sensitivity at 0.1 V is the lowest (due to the nearly 0 slope), while that for a gate voltage in the close neighbor of the "Dirac" point should be the highest (because of the steepest slope). When the embryonic chicken cardiomyocyte is electrically activated, the transistor is able to detect the conductance change and record the neural activity, as demonstrated in Fig. 8.6c. The magnification of the recorded action potentials at various gate voltages are shown in Fig. 8.6d, where the blue, green, and red color represent that the graphene is hole-dominant, neutral (at "Dirac" point), and electron-

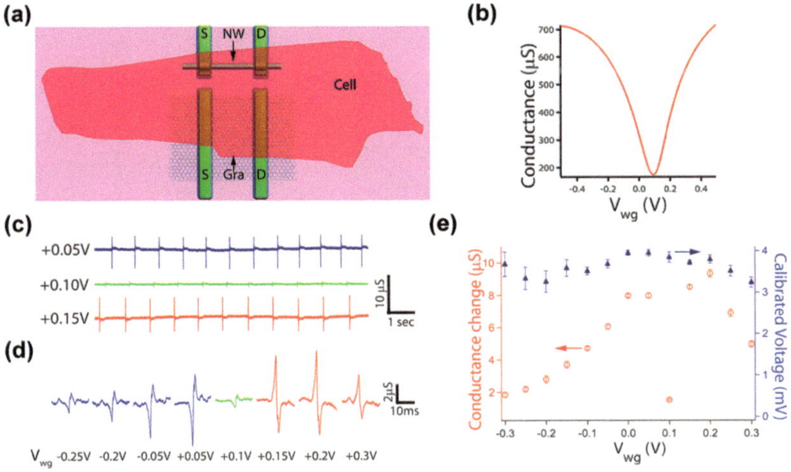

Figure 8.6 (a) Illustrations of a nanowire (NW) transistor and a graphene (Gra) transistor used for neuronal recording. (b) Measured conductance of graphene as a function of the gate voltage (V_{wg}) in water. (c) Recorded activities of cardiomyocytes by a graphene transistor when the gate was biased at different voltages (+0.05 V, +0.1 V, and +0.15 V). The corresponding sensitivities are for these gate voltages are 2020, 398, and 2290 μS/V, respectively. (d) Magnification of the recorded action potentials at various gate voltages, where the blue, green, and red color represent that the graphene is hole-dominant, neutral (at "Dirac" point), and electron-dominant, respectively. (e) Measured conductance change of graphene and the calibrated extracellular voltage as a function of V_{wg}. Reprinted with permission from [Cohen-Karni et al. (2010)] Copyright American Chemical Society.

dominant, respectively. As predicted earlier, the sensitivity of the device, represented by the amplitude of the recorded signal, is nearly the lowest near 0.1 V, but it is very high when V_{wg} is in the close vicinity of the "Dirac" point (e.g., 0.05 V, 0.15 V, 0.2 V, etc.). It is interesting to note that the recorded signal in the hole-dominant region is opposite to that in the electron-dominant region, due to the opposite sign of the slope in Fig. 8.6b. Figure 8.6e summarizes the measured conductance change of graphene (red open circles) as a function of V_{wg}, which confirms that the sensitivity is related to the slope of the conductance in Fig. 8.6b. The constant calibrated extracellular voltage indicated by the blue triangles in

Fig. 8.6e suggests that this method is reliable for neural activity recording.

8.4 Fast Electronic Devices

In addition to utilize the electrons in graphene to probe the ionic and electrical activities in solution, the high dielectric constant (κ) of the liquid can be utilized to improve the mobility of the charge carriers in graphene [Chen et al. (2009a,b); Farmer et al. (2009); Ponomarenko et al. (2009); Xia et al. (2010b)], due to the dielectric screening effect (section 5.1.2).

In comparison with the liquid-gate configuration, here the liquid is placed on the top of a bottom-gate graphene transistor, as

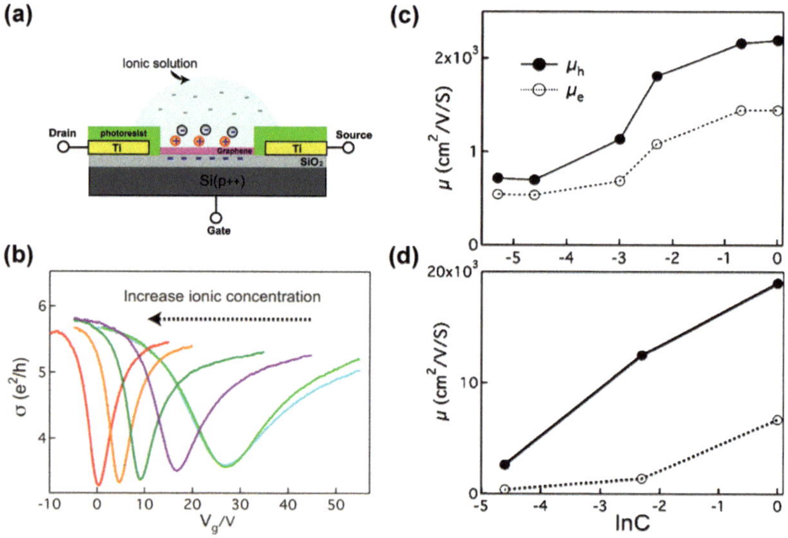

Figure 8.7 (a) Schematics of a bottom-gate graphene transistor where a solution is placed on top. (b) Measured sheet conductance of graphene in aqueous solutions which contain NaF of different concentrations (from right to left: 5 mM, 10 mM, 0.05 M, 0.1 M, 0.5 M, 1 M). (c) The hole and electron mobility (μ_h and μ_e) as a function of the electrolyte concentration (C, shown in a logarithmic scale). (d) A greater enhancement of the mobility obtained by a different device. Reprinted with permission from [Chen et al. (2009b)] Copyright American Chemical Society.

illustrated in Fig. 8.7a. This is because the transistor channel needs to be immersed in a high-κ environment, and the method to apply the gate potential is not critically important. Figure 8.7b shows the measured sheet conductance of graphene (σ) in aqueous solutions which contain NaF of different concentrations. As the ionic concentration increases (from 5 mM to 1 M), the "Dirac" point (the minimum of the conductance) moves towards left, and the slope of the conductance near the "Dirac" point becomes steeper. Since the net charge carrier density in graphene is defined by $C_{\text{ox}}V_g$, where C_{ox} is the capacitance of the oxide layer in the bottom gate (constant), the derivative of σ over V_g then represents the mobility (μ), because of $\frac{\mathrm{d}\sigma}{\mathrm{d}V_g} = C_{\text{ox}}\frac{\mathrm{d}\sigma}{\mathrm{d}en_{\text{net}}} = C_{\text{ox}}\mu$.

Figure 8.7c and Table 8.1 summarize the enhancement of the charge carrier mobility as a function of the electrolyte concentration. μ increases with the concentration of NaF, as discussed earlier. μ could be further improved depending on the design of the electronic device (Fig. 8.7d) or the choice of the solution. This indicates that liquids, as high-κ environment, can be used to design next generation electronic devices, as the high charge carrier mobility is usually desired. It can also be seen from these graphs that the difference between the electron and hole mobility increases. This is consistent with the observation by liquid gated Hall effect measurement, as presented in the previous section, which should be taken into account in the future development of the fast electronic devices.

Table 8.1 Summary of the mobility obtained in Fig. 8.7b. Reprinted with permission from [Chen et al. (2009b)] Copyright American Chemical Society

	μ_{h} (cm^2/(V·s))	μ_{e} (cm^2/(V·s))
argon	118	261
5 mM	716	543
10 mM	698	535
0.05 M	1133	688
0.1 M	1808	1083
0.5 M	2158	1442
1 M	2192	1442

It is worth to mention that the slope of σ over V_g is lower when the charge carrier density is high. Take the red curve shown in Fig. 8.7b as an example, the slope near $V_g = -9$ V is nearly 0. It indicates that the mobility of graphene with high charge carrier density in this device is low, also because of the dielectric screening. This is possibly due to the reduced screening of the impurity potential by conduction electrons, a study confirms that the carrier-density independent conductivity limited by the short-range scattering could be decreased by 38% [Jang et al. (2008)]. An investigation using angle-resolved photoemission spectroscopy (section 6.2.4) further explains that the renormalization of the "Dirac" cone due to the screening of the electron-electron interaction could be related to the dielectric screening phenomenon [Siegel et al. (2013)].

8.5 Outlook

In essence, the linear gap-less band structure and the atomic thickness of graphene (section 3.1.2) gives several unparalleled advantages to the charge carriers (i.e., electrons or holes). (i) The massless Fermions moves very fast in response to an electric field, namely the mobility is very high. (ii) The gap-less band structure delivers a smooth transition between the electron- and hole-dominant regions, where a conductor with the mixture of electron-hole gas is also possible to obtain. This allows the charge carriers to respond to the environment near the "Dirac" point, and hence more information of the interaction (between the charge carriers in graphene with the environment) could be extracted. (iii) The linear density-of-states enables the observation of the V-shaped sheet conductance (proportional to the total carrier density, as shown in Fig. 3.5) and quantum capacitance (Fig. 3.6). This indicates that a small change in the electrical properties of graphene in response to the gate voltage could be observable, in comparison with the conventional semiconductors (Fig. 2.14). (iv) The atomic thickness of graphene makes it very sensitive to the environment and renders itself as an ideal candidate for low-dimensional electronic devices.

Most of the applications described in this chapter therefore involve the use of monolayer or few-layered graphene in electrolytes (except for section 8.1.1). These applications either utilize the electrons in graphene to probe the ionic or electrical activities in solution, or take advantage of the liquid to enhance the performance of the electronic devices. However, the electronic properties of multi-layered graphene membrane at the graphene–electrolyte interfaces are not widely studied. These systems are intensively investigated for the applications on energy storage, water treatment, and neuron stimulation. These topics are very interesting for the global society. Future studies in these fields should be focused on the ionic properties at graphene–electrolyte interfaces.

Appendix A

Moments of Boltzmann Transport Equation: The Fluid Equations

A.1 Dyadic Tensor

In physics, tensors are objects that used to describe physical quantities. An nth order tensor is represented by an array where its dimensionality is n.

By convention, tensors with various orders have different names. A scalar is a 0th order tensor which is single number, such as temperature. A vector is a 1st order tensor which is denoted by a one-dimensional array, such as position. A dyadic is a 2nd order tensor which is represented by a two-dimensional array (matrix), an example of a physical dyadic is stress.

A 2nd order tensor $\overleftrightarrow{T}$ as the dyadic product of two vectors $\vec{A}$ and $\vec{B}$ can be written in algebraic form as

$$\overleftrightarrow{T} = \vec{A}\vec{B} = \begin{pmatrix} A_i B_i & A_i B_j & A_i B_k \\ A_j B_i & A_j B_j & A_j B_k \\ A_k B_i & A_k B_j & A_k B_k \end{pmatrix}, \tag{A.1}$$

where $\vec{A} = A_i\hat{e}_i + A_j\hat{e}_j + A_k\hat{e}_k$ and $\vec{B} = B_i\hat{e}_i + B_j\hat{e}_j + B_k\hat{e}_k$. $\hat{e}_i$, $\hat{e}_j$, and $\hat{e}_k$ are the standard bases in the vector space.

For the derivation of the fluid equations using BTE, it is useful to review some of the properties of dyadics.

Graphene–Electrolyte Interfaces: Electronic Properties and Applications
Hualin Zhan

ISBN 978-981-4774-94-9 (Hardcover), 978-1-003-04487-1 (eBook)
www.jennystanford.com

1. *The dyadic product of tensors is distributive,*

$$\vec{A}(\vec{B}+\vec{C}) = \vec{A}\vec{B} + \vec{A}\vec{C}, \tag{A.2a}$$

$$(\vec{A}+\vec{B})\vec{C} = \vec{A}\vec{C} + \vec{B}\vec{C}. \tag{A.2b}$$

2. *Scalar multiplication of a dyadic,*

$$f(\vec{A}\vec{B}) = (f\vec{A})\vec{B} = \vec{A}(f\vec{B}), \tag{A.3}$$

   where f is a scalar.

3. *Product of a vector and a dyadic,*

   3.1 Dot product,

$$\vec{A}\cdot(\vec{B}\vec{C}) = (\vec{A}\cdot\vec{B})\vec{C}, \tag{A.4a}$$

$$(\vec{A}\vec{B})\cdot\vec{C} = \vec{A}(\vec{B}\cdot\vec{C}). \tag{A.4b}$$

   3.2 Cross product,

$$\vec{A}\times(\vec{B}\vec{C}) = (\vec{A}\times\vec{B})\vec{C}, \tag{A.5a}$$

$$(\vec{A}\vec{B})\times\vec{C} = \vec{A}(\vec{B}\times\vec{C}). \tag{A.5b}$$

The multiplication of dyadic tensors are more complicated and they are not used in this book. Readers who are interested in this topic can find more details in tensor analysis textbooks [Lipschutz et al. (2009)].

In vector analysis, divergence is an important concept which is used to describe the flow flux of a vector field in an infinitesimal volume. When analyzing the collective behavior of a group of particles (such as liquids, gases, or plasmas), it is useful to find the divergence of a tensor field. This requires us to have a gradient operator (∇) acting on a dyadic. We must not forget that ∇ is not only a vector but also a differential operator. Therefore, similar to the divergence in vector analysis where $\nabla\cdot(f\vec{A}) = (\nabla f)\cdot\vec{A} + f\nabla\cdot\vec{A}$, we have

$$\begin{aligned}\nabla\cdot(\vec{A}\vec{B}f) &= \nabla\cdot(f\vec{A}\vec{B})\\ &= [\nabla\cdot(f\vec{A})]\vec{B} + (f\vec{A}\cdot\nabla)\vec{B}\\ &= (\nabla f\cdot\vec{A})\vec{B} + (\nabla\cdot\vec{A})f\vec{B} + (f\vec{A}\cdot\nabla)\vec{B}.\end{aligned} \tag{A.6}$$

This relation is particularly useful when deriving the kth order moment of BTE.

A.2 $\mathbf{k^{th}}$ Order Moment

Taking the moments of BTE is a general analytical technique for theoretical studies of the collective behavior of particles. A kth order moment is defined as the integration of the product of BTE and $\vec{v}^k$ $(\underbrace{\vec{v}\vec{v}\cdots\vec{v}}_{k}$, a kth order tensor which is a dyadic product of k vectors $\vec{v}$) over the velocity (momentum) space

$$\int \left(\frac{\partial f}{\partial t} + \vec{v} \cdot \nabla f + \frac{\vec{F}}{m} \cdot \nabla_v f \right) \vec{v}^k \mathrm{d}\vec{v} = \int \left[\left(\frac{\partial f}{\partial t} \right)_c \right] \vec{v}^k \mathrm{d}\vec{v}, \quad \text{(A.7)}$$

where $\vec{v}$ is the velocity of a particle, and f is the distribution function treated as a scalar. The general solution is complicated, it is then useful to perform integration for each term in the equation.

The first term on the left-hand side (LHS):

$$\int \left(\frac{\partial f}{\partial t} \right) \vec{v}^k \mathrm{d}\vec{v} = \frac{\partial}{\partial t} n \langle \vec{v}^k \rangle - n \left\langle \frac{\partial \vec{v}^k}{\partial t} \right\rangle, \quad \text{(A.8)}$$

where a relation $\int f \mathrm{d}\vec{v} = n$ is used, and n is the number density of particles. $\langle x \rangle = \frac{1}{n} \int f x \mathrm{d}\vec{v}$ represents the average value of a quantity x over velocity space.

For the second term on the LHS:

$$\begin{aligned} \int (\vec{v} \cdot \nabla f) \vec{v}^k \mathrm{d}\vec{v} &= \nabla \cdot (n \langle \vec{v}\vec{v}^k \rangle) - n \langle (\nabla \cdot \vec{v}) \vec{v}^k \rangle - n \langle (\vec{v} \cdot \nabla) \vec{v}^k \rangle \\ &= \nabla \cdot (n \langle \vec{v}\vec{v}^k \rangle) - n \langle \vec{v} \cdot \nabla \vec{v}^k \rangle, \end{aligned} \quad \text{(A.9)}$$

where a relation $\nabla \cdot \vec{v} = 0$ is used as position and velocity are independent to each other in phase space.

The third term on the LHS then becomes:

$$\begin{aligned} \int \left(\frac{\vec{F}}{m} \cdot \nabla_v f \right) \vec{v}^k \mathrm{d}\vec{v} &= -n \left\langle \left(\nabla_v \cdot \frac{\vec{F}}{m} \right) \vec{v}^k \right\rangle - n \left\langle \left(\frac{\vec{F}}{m} \cdot \nabla_v \right) \vec{v}^k \right\rangle \\ &= -n \left\langle \left(\frac{\vec{F}}{m} \cdot \nabla_v \right) \vec{v}^k \right\rangle. \end{aligned} \quad \text{(A.10)}$$

For a system where the collision processes are insignificant, the right-hand side (RHS) is 0 as the distribution function does not change over time. It should be noted that for the study of transport phenomena, the collision term should be considered and it has different forms of expression (e.g., BGK, Boltzmann collision integration, Fokker-Planck, etc.) and so the RHS is non-zero.

A.3 The Fluid Equations

Fluid equations study the dynamical behavior of a group of particles (fluid parcel) in position space, which can be obtained by taking the moments of BTE. Note that the collision term $\left(\frac{\partial f}{\partial t}\right)_c$ is neglected in the following derivation.

A.3.1 Continuity Equation

The 0th order moment gives the continuity (or mass transport) equation. Substitution of $k = 0$ into Eqs. (A.7–A.10) yields

$$\frac{\partial n}{\partial t} + \nabla \cdot (n\vec{u}) = 0, \tag{A.11}$$

where $\langle \vec{v} \rangle = \vec{u}$ is the mean velocity of a fluid parcel.

A.3.2 Momentum Conservation

The 1st order moment ($k = 1$) gives the momentum conservation equation,

$$\frac{\partial}{\partial t}(n\vec{u}) + \nabla \cdot (n\langle \vec{v}\vec{v} \rangle) - n\left\langle \frac{\vec{F}}{m} \right\rangle = 0. \tag{A.12}$$

It is convenient to write $\vec{v}$ as a function of the mean velocity $\vec{u}$ and a perturbation term $\vec{w}$, $\vec{v} = \vec{u} + \vec{w}$ where $\langle \vec{u} \rangle = \vec{u}$, $\langle \vec{w} \rangle = 0$, and $\langle \vec{w}\vec{w} \rangle = \frac{1}{nm}\overleftrightarrow{P}$. $\overleftrightarrow{P} = p\overleftrightarrow{I} - \Pi_v - \Pi_s$ is a general stress tensor which is associated with pressure p, viscosity (Π_v), and strain rate tensor (Π_s), where $\overleftrightarrow{I}$ is the unit tensor. Readers who are interested in the derivation of the general stress tensor are recommended to read fluid (or plasma) theory textbooks [Boyd and Sanderson (2003)]. Equation (A.12) can be re-written as

$$nm\left[\frac{\partial \vec{u}}{\partial t} + (\vec{u} \cdot \nabla)\vec{u}\right] = -\nabla p + \Pi_v + \Pi_s + n\vec{F}. \tag{A.13}$$

This is the famous Navier-Stokes (or Euler) equation.

A.3.3 Energy Conservation

As the order of moment increases, the workload of the calculation escalates very quickly. 2nd order moment ($k = 2$) gives the energy conservation equation,

$$n\frac{\mathrm{d}}{\mathrm{d}t}\left(\frac{3}{2}k_{\mathrm{B}}T + \frac{1}{2}mu^2\right) = -\nabla \cdot (\vec{Q} + \overleftrightarrow{P} \cdot \vec{u}) + n\vec{F} \cdot \vec{u}, \qquad \text{(A.14)}$$

where $\vec{Q} = \frac{1}{2}nmw^2\vec{w}$ is the heat flux. Note that an approximation $\vec{w}\vec{w} \approx w^2$ and a relation $\frac{3}{2}nk_{\mathrm{B}}T = \frac{1}{2}nm\langle w^2\rangle$ are used in the derivation.

Appendix B

Electric Field in the Compact Stern Layer

The mPB theory is derived by modifying the Boltzmann distribution function, and then substituting it into the Poisson's equation (Eq. 4.51). This is a different approach to account for the steric effects than the conventional Poisson–Nernst–Planck equation or the vibration model. Therefore, the mPB theory has been adopted here to verify the calculated electric field in section 4.3.3.1 from a different perspective. It is worth noting that, similar to the mPB theory when $\gamma = 1$, the volumetric density ratio of cations and anions can still be described by Boltzmann term $\frac{n_{c_i}}{n_{a_i}} = \exp(-2u)$, even for very high surface concentration [Oldham (2008)]. It is therefore straightforward to obtain the ionic concentrations

$$n_{c_i} = \frac{n_{t_i} \exp(-u)}{2\cosh(u)}, \tag{B.1a}$$

$$n_{a_i} = \frac{n_{t_i} \exp(u)}{2\cosh(u)}, \tag{B.1b}$$

where $u = \frac{ze\varphi}{k_B T}$, $n_{t_i} = n_{c_i} + n_{a_i}$ is the total volumetric density for both cations and anions in sublayer i. The indices c and a denote cations and anions, respectively. Hence, the volumetric net charge density can be written as

$$\rho_{\text{net}} = \frac{z_c e n_{t_i}}{2} \cdot \frac{\exp(-u) - \exp(u)}{\cosh(u)} = -\rho_t \tanh(u), \tag{B.2}$$

Graphene–Electrolyte Interfaces: Electronic Properties and Applications
Hualin Zhan

ISBN 978-981-4774-94-9 (Hardcover), 978-1-003-04487-1 (eBook)
www.jennystanford.com

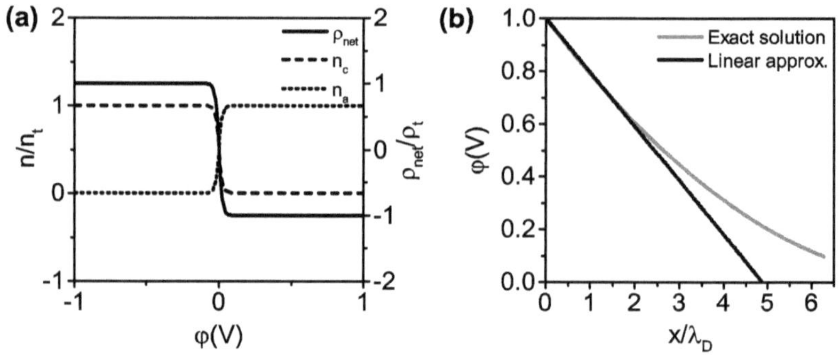

Figure B.1 (a) Modified Boltzmann distribution of n_c (dashed), n_a (dotted), and ρ_{net} (solid). (b) Comparison of the exact solution for Eq. B.3 and its linear approximation.

where $\rho_t = \frac{z_c e n_{t_i}}{2}$. Distribution of n_c, n_a, and ρ_{net} have been plotted in Fig. B.1b. It should be noted that at high bias voltage, the net charge density in the whole compact layer is very low comparing with that in bulk electrolyte at the same voltage (if possible), because the denominator $\cosh(u)$ rises very quickly as φ increases.

Substituting the Eq. B.2 into Poisson's equation gives

$$\frac{d\varphi}{dx} = \mp\frac{k_{\text{B}}T}{ze\lambda_{\text{D}}}\Theta, \tag{B.3}$$

where $\Theta = \sqrt{2\ln[\cosh(u)]}$, $\lambda_{\text{D}} = \sqrt{\frac{k_{\text{B}}T\varepsilon}{z^2e^2n_{t_i}}}$ is the Debye length, ε is the permittivity, z is the ionic charge, and e is electron charge. Here the upper/bottom signs in equations are for positively/negatively biased electrodes, respectively. When the electrode is biased at high voltage, $\pm u \gg 1$ and $e^{\mp u} \to 0$ (for example, $e^{-u} = 3.64 \times 10^{-4}$ when $\varphi = 0.2V$).

Therefore, $\Theta = \sqrt{2(\pm u - \ln 2)} = \sqrt{2(|u| - \ln 2)}$, since $\cosh(u) = \frac{e^u + e^{-u}}{2} \approx \frac{e^{\pm u}}{2}$. Thus the solution of Eq. B.3 can be obtained as

$$u = \pm\left[\frac{1}{2}\left(c - \frac{x}{\lambda_{\text{D}}}\right)^2 + \ln 2\right], \tag{B.4}$$

where $c = \sqrt{2(|u_s| - \ln 2)}$ is a constant determined by letting $u(x = 0) = |u_s| = \frac{|ze\varphi^m|}{k_{\text{B}}T}$. Note that this relation is only valid when $|u|$ is large, namely $\dfrac{x}{\lambda_{\mathrm{D}}}$ is small. The electric field strength at position x_i inside the compact layer is

$$E_i = -\left(\frac{d\varphi}{dx}\right)_{x_i} = \pm\frac{k_{\mathrm{B}}T}{ze\lambda_{\mathrm{D}}}\left(c - \frac{x_i}{\lambda_{\mathrm{D}}}\right). \tag{B.5}$$

Substituting this relation into Eq. C.5 gives the same result which can be derived from Eq. B.1 when $|u| \gg 1$. This indicates these two theories are consistent at high potential.

If $\frac{x_i}{\lambda_{\mathrm{D}}} \ll c$ (usually true when the electrode is biased at high voltage), $E_i \approx \pm\frac{k_{\mathrm{B}}T}{ze\lambda_{\mathrm{D}}}c$ is a constant. For example, when $\varphi^m = 1$ V, $c \approx 8.8$ and $\frac{x_i}{\lambda_{\mathrm{D}}} \sim \frac{r_{\mathrm{H}}}{\lambda_{\mathrm{D}}} \in (0, 2)$ (depending on the choice of r_{H}).

The exact solution (obtained numerically) for Eq. B.3 and its linear approximation (derived from Eq. B.5) are compared in Fig. B.1(b). It can be seen from the figure that the linear approximation is able to describe the exact solution when $\frac{x}{\lambda_{\mathrm{D}}}$ is small, hence the electric field in compact layer is constant when the electrode is biased at high voltage. This indicates the solution obtained by Eq. 4.53 is consistent with the mPB theory, if the approximations $r_x \ll c\lambda_{\mathrm{D}}$ and $\left|\frac{ze\varphi}{k_{\mathrm{B}}T}\right| \gg 1$ are applied. It is concluded that the ambipolar diffusion theory can be used to investigate the compact Stern layer. However, the sharp peaks appeared in Fig. 4.15b indicate further study of the electrical double layer structure is needed.

Appendix C

The Poisson–Nernst–Planck Equation for the Compact Stern Layer

The Poisson–Nernst–Planck equation can be derived by calculating and re-arranging the 1st-order moment of Boltzmann transport equation [Boyd and Sanderson (2003); Zhan et al. (2016); Zhan and Hu (2011)], regardless the exact expression of distribution function. It has been used to describe particle motion in many different states of matter, such as liquid [Bard and Faulkner (2000)], solid [Kittel (2004)], and plasmas [Boyd and Sanderson (2003)], in spite of their different forms of distribution functions.

C.1 General Form to Find Potential Distribution

In general, when a system is at equilibrium, $\Gamma_c = \Gamma_a$ (or $e(z_c\Gamma_c + z_a\Gamma_a) = 0$) in the compact layer [Boyd and Sanderson (2003); Zhan et al. (2016)], where $\Gamma_c = -D_c\nabla n_c + n_c b_c E$ and $\Gamma_a = -D_a\nabla n_a + n_a b_a E$ are the fluxes, D_c and D_a are the diffusion coefficients, n_c and n_a are the volumetric ionic density, for cations and anions, respectively. Therefore, to achieve a flux balance between sublayers i and $i+1$ in the compact Stern layer, we obtain

$$(e(z_c\Gamma_c + z_a\Gamma_a))_i = (e(z_c\Gamma_c + z_a\Gamma_a))_{i+1}. \tag{C.1}$$

Graphene–Electrolyte Interfaces: Electronic Properties and Applications
Hualin Zhan

ISBN 978-981-4774-94-9 (Hardcover), 978-1-003-04487-1 (eBook)
www.jennystanford.com

Re-arranging equation (C.1) gives

$$\begin{aligned}\sigma_i E_i - \sigma_{i+1} E_{i+1} &= \mathcal{X}_i \\ &= z_c e D_c (\nabla n_{c_i} - \nabla n_{c_{i+1}}) \\ &\quad + z_a e D_a (\nabla n_{a_i} - \nabla n_{a_{i+1}}).\end{aligned} \tag{C.2}$$

For all the sublayers, a general form of the system of equations for finding the potential distribution can be obtained

$$\begin{pmatrix} \sigma_2 & -\sigma_3 \\ w_2 & w_3 \end{pmatrix} \cdot \begin{pmatrix} E_2 \\ E_3 \end{pmatrix} = \begin{pmatrix} \mathcal{X}_2 \\ \varphi^m - \varphi_2^{\text{lim}} \end{pmatrix}, \tag{C.3}$$

where $\mathcal{X}_2 = z_c e D_c (\nabla n_{c_2} - \nabla n_{c_3}) + z_a e D_a (\nabla n_{a_2} - \nabla n_{a_3})$.

As shown in Fig. B.1b, when the electrode is biased at high voltage and $x_i/\lambda_D \ll c$ in Eq. B.5, $\nabla n_{c_i} - \nabla n_{c_{i+1}} = 0$ because $\nabla n_{c_i} = \frac{dn_{c_i}}{d\varphi}\frac{d\varphi}{dx} \approx 0$ at high voltage. Similarly, we can obtain $\nabla n_{a_i} - \nabla n_{a_{i+1}} = 0$. Hence equation system (C.3) reduces to Eq. 4.53 in section 4.3.3.1.

C.2 Net Charge Density in Sublayer *i*

When electrode is biased at low voltage, the assumptions made in section C.1 are invalid, therefore the ionic charge density need to be included in Eqs. C.1 and C.3.

The areal net charge density in sublayer i is given by Gauss's law

$$\rho_{\text{net}_i}^{\text{2D}} = E_{i-1} - E_{i+1}. \tag{C.4}$$

E_1 represents the electric field in the diffuse layer which is close to the interface between the whole compact Stern layer and the diffuse layer. And it can be obtained by solving Poisson-Boltzmann equation (Eq. 4.42). If sublayer i is right next to the electrode surface, $\rho_{\text{net}_i}^{\text{2D}} = 0$ (excluding the charge on electrode) due to the definition of Stern layer.

If sublayer width w_i is small enough, the volumetric net charge density can be written as $\rho_{\text{net}_i} = \rho_{\text{net}_i}^{\text{2D}}/w_i$. Therefore, the volumetric ionic density can be expressed as

$$n_{c_i} = \frac{1}{2}\left(n_{t_i} + \frac{E_{i-1} - E_{i+1}}{z_c e w_i}\right), \tag{C.5a}$$

$$n_{a_i} = \frac{1}{2}\left(n_{t_i} - \frac{E_{i-1} - E_{i+1}}{z_c e w_i}\right). \tag{C.5b}$$

Substituting this system of equations into Eqs. C.2 and D.1 gives the complete form which allows us to find the potential distribution in the whole compact layer. Note that this is a non-linear system and to find its exact solution, we need to use Steepest Descent and Newton-Raphson methods.

It should be noted that this set of equations can not be used to estimate the ionic density if the aforementioned assumptions in section C.1 were applied. The reason is that $z_c e w_i$ will be much smaller than $E_{i-1} - E_{i+1}$ despite of $E_{i-1} - E_{i+1} \approx 0$.

Appendix D

Conductivities of Sublayer *i* in the Compact Stern Layer

When the actual ionic areal density exceeds its limit, all the ions at the electrode surface are arranged in such a way that the ions right next to the electrode surface form compact layer(s), which is the so-called compact Stern layer. Ideally we have all ions arranged compactly, like those in solid, in the compact Stern layer. If there are enough ions in the compact Stern layer, we will have a few separate layer (sublayer) of ions in it. Here each sublayer inside the whole compact layer is denoted as sublayer i ($i \geq 2$), where sublayer 1 represents the diffuse layer. The width of each sublayer, w_i, is roughly equal to the radius/diameter of the ion (i.e., w_2 and w_3 in section 4.3.3.1).

Since the ions are still envisaged as solid spheres with fixed radius, it is, however, impossible to have all the ions compactly arranged in every sublayer in the compact Stern layer. Because the number of ions is not always exactly the same as the total number of states available in the compact Stern layer (which is similar to atomic layers in solid). Therefore, we keep the ions in the sublayers close to the electrode compactly arranged (i.e., $w_3 = r_{\mathrm{H}}$), but not the sublayer at the interface of this compact Stern layer and bulk solution (i.e., $w_2 = r_x - r_{\mathrm{H}}$).

Graphene–Electrolyte Interfaces: Electronic Properties and Applications
Hualin Zhan

ISBN 978-981-4774-94-9 (Hardcover), 978-1-003-04487-1 (eBook)
www.jennystanford.com

The conductivities of these sublayers directly affect the result of the electric field strength near the electrode surface, hence the differential capacitance, as can be seen in Eq. 4.53 in section 4.3.3.1. Therefore determining these conductivities is essential.

D.1 Mobile Ions

If all the ions stacked in the whole compact layer are still mobile (due to vibration, for example), then both cations and anions will contribute to the total current (classical electrolyte conductance model [Atkins and de Paula (2006)]). For symmetric electrolyte ($z_c = -z_a$), we have

$$\begin{aligned} \sigma_i &= w_i^{-1}(n_{c_i}^{2\mathrm{D}} z_c e b_c + n_{a_i}^{2\mathrm{D}} z_a e b_a) \\ &= w_i^{-1}[n_{t_i}^{2\mathrm{D}} z_c e b_c + n_{a_i}^{2\mathrm{D}} z_a e(b_c + b_a)] \\ &\overset{\text{or}}{=} w_i^{-1}[n_{t_i}^{2\mathrm{D}} z_a e b_a + n_{c_i}^{2\mathrm{D}} z_c e(b_c + b_a)], \end{aligned}$$

where σ_i is the conductivity, w_i is the sublayer width, $n_{t_i}^{2\mathrm{D}} = n_{c_i}^{2\mathrm{D}} + n_{a_i}^{2\mathrm{D}}$ is the total areal density of both cations and anions, $n_{c_i}^{2\mathrm{D}}$ and $n_{a_i}^{2\mathrm{D}}$ are the areal density, z_c and z_a are the charge, b_c and b_a are the mobility, for cations and anions in sublayer i, respectively.

It can be seen from this equation that when $n_{a_i}^{2\mathrm{D}} = 0$, $n_{c_i}^{2\mathrm{D}} = 0$, or $b_c = -b_a$, the second term in square brackets at the right hand side is 0. Hence σ_i is dependent on $n_{t_i}^{2\mathrm{D}}$. As shown in Fig. B.1b, generally we have $n_{a_i}^{2\mathrm{D}} \approx 0$ or $n_{c_i}^{2\mathrm{D}} \approx 0$ when the electrode is biased at high voltage. Since $n_{t_i}^{2\mathrm{D}}/w_i$ is the same in each sublayer, σ_i will be the same as well, i.e. $\sigma_2 = \sigma_3$ (see section 4.3.3.1). It should be noted that $n_{a_i}^{2\mathrm{D}} \neq 0$, $n_{c_i}^{2\mathrm{D}} \neq 0$, and $b_c \neq -b_a$ at low bias voltages, thus Eq. D.1, along with the ionic charge equation in section D.2, should be substituted into Eq. 4.53 in section 4.3.3.1 to obtain the potential distribution.

D.2 Immobile Ions

In the case that all the ions in the whole compact layer are so compactly arranged at very high voltage so that they could not move too much, then the electrons/holes in anions/cations adsorbed on positive/negative electrode are the charge carriers (analogy to intrinsic semiconductor). Therefore the conductivity of each sublayer will be determined by the net charge at the sublayer interfaces,

$$\sigma_i^* = \frac{e}{w_i}(n_{c_i}^{2D} z_c b_h + n_{a_i}^{2D} z_a b_e) = \frac{e z_a b_e n_{t_i}^{2D}}{w_i}, \tag{D.1}$$

where $b_h = -b_e$, b_h and b_e are the hole/electron mobility in cations/anions. It can be seen from the equation that σ_i^* is the same in every sublayer due to the reasons as mentioned before.

Appendix E

Constant Variables in the Compact Stern Layer at High Potential

Total ionic areal density within the compact Stern layer and on OHP can be calculated as follows,

$$
\begin{aligned}
n_t^{2\mathrm{D}} &= \int_{x_2}^{x_1} n_t dx = \int_{\varphi_2}^{\varphi_1} \frac{n_t}{d\varphi/dx} dx \\
&= -\sqrt{\frac{2n_0 \varepsilon k_\mathrm{B} T}{z^2 e^2}} \int_{u_2/2}^{u_1/2} \frac{\cosh(u)}{\sinh(u/2)} d\left(\frac{u}{2}\right) \\
&= -\sqrt{\frac{2n_0 \varepsilon k_\mathrm{B} T}{z^2 e^2}} \int_{u_2/2}^{u_1/2} \frac{2\sinh^2(u)+1}{\sinh(u/2)} d\left(\frac{u}{2}\right) \\
&= -\sqrt{\frac{2n_0 \varepsilon k_\mathrm{B} T}{z^2 e^2}} \left[2\Delta \cosh\left(\frac{u}{2}\right) - \frac{x_1 - x_2}{\lambda_\mathrm{D}} \right],
\end{aligned}
$$

where $\Delta \cosh\left(\frac{u}{2}\right) = \cosh\left(\frac{u_1}{2}\right) - \cosh\left(\frac{u_2}{2}\right)$, and the relation $\ln \frac{|\tanh(u_1/4)|}{|\tanh(u_2/4)|} = -\frac{x_1 - x_2}{\lambda_\mathrm{D}}$ has been used [Bard and Faulkner (2000)]. At high potential, the total ionic areal density at OHP is a constant ($n_\mathrm{OHP}^{2\mathrm{D}} = n_\mathrm{max}^{2\mathrm{D}}$). Since $\frac{x_1 - x_2}{\lambda_\mathrm{D}}$ is constant, hence $\Delta \cosh\left(\frac{u}{2}\right)$ is a constant. Because $\cosh(u)$ is a monotonic function when $u > 0$ or $u < 0$, u_2 is also a constant.

Graphene–Electrolyte Interfaces: Electronic Properties and Applications
Hualin Zhan

ISBN 978-981-4774-94-9 (Hardcover), 978-1-003-04487-1 (eBook)
www.jennystanford.com

Therefore, when the total ionic density at OHP reaches the maximum ($n^{2D}_{\max}$), φ_2 (or $\varphi_2^{\lim}$) is a constant. Since $C_D^{\lim}$ in section 4.3.3.1 has only one variable $\varphi_2^{\lim}$, therefore it is also a constant.

As mentioned before, n^{2D}_{OHP} rises as the electrode potential increases. At one point when the electrode potential approaches the value $(\varphi^m)^{\lim}$, n^{2D}_{OHP} reaches $n^{2D}_{\max}$. Then the distance between x_2 and 0 is still r_{H}, although it will continue increase with the electrode potential. Hence $S^{\lim}$ from section 4.3.3.1 can be written

$$S^{\lim} = \frac{(\varphi^m)^{\lim} - (\varphi^m)^{\lim}}{r_{\mathrm{H}}}. \tag{E.1}$$

It can be seen that $S^{\lim}$ is a constant as well. This relation is useful for numerical calculation.

Appendix F

Fabrication Process of a Graphene Hall Device

The graphene/SiO_2/Si samples can either be commercially purchased or grown by the methods discussed in section 6.1. To reduce the time that the graphene was in contact with chemicals during the fabrication, a fabrication process using an image reversal photoresist (TI35ES) could be used. At first, the graphene/SiO_2/Si sample was annealed at 150°C in vacuum for 1h to improve the adhesion between graphene and the substrate. A photolithography process was then used to obtain the graphene Hall structures using a TI35ES as the positive photoresist. A subsequent lithography was then conducted to produce the electrode pattern for the Hall structures using a TI35ES as the negative photoresist. This was followed by a metal evaporation step to deposit the contacts. As the Hall measurements were performed in solution, the sample surface (except for the graphene Hall structures and the electrical contacts on the right) was covered through a lithography process using a bio-compatible epoxy-like photoresist (SU-8). To minimize defects, the samples were annealed at 150°C in vacuum. They were then mounted on printed circuit boards with those contacts wire-bonded. Before the measurements, all the contacts except for the Hall structures (including the wire bonds) were covered by epoxy. Figure F.1 illustrates the fabrication process, where the photograph

Graphene–Electrolyte Interfaces: Electronic Properties and Applications
Hualin Zhan

ISBN 978-981-4774-94-9 (Hardcover), 978-1-003-04487-1 (eBook)
www.jennystanford.com

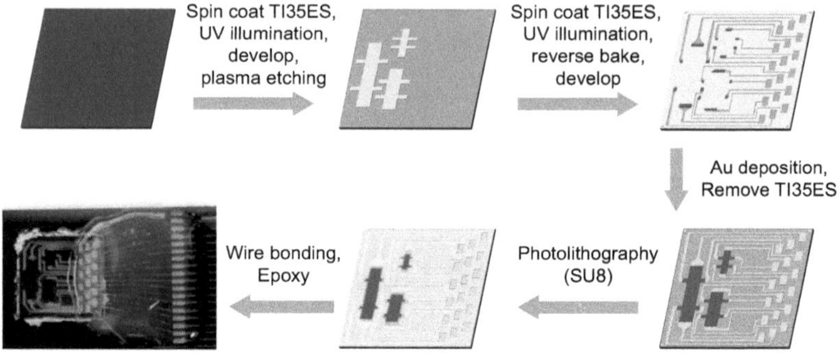

Figure F.1 Fabrication process of graphene Hall structures, where the photograph shows a fabricated device with all contacts covered by epoxy or SU-8 except for the graphene region.

shows a fabricated device with all contacts covered by epoxy or SU-8 except for the graphene region.

References

Abajo, F. J. G. d. (2013). Multiple excitation of confined graphene plasmons by single free electrons, *ACS Nano* **7**, 12, pp. 11409–11419

Adam, S., Hwang, E. H., Galitski, V. M. and Das Sarma, S. (2007). A self-consistent theory for graphene transport, *Proc Natl Acad Sci U S A* **104**, 47, pp. 18392–18397, doi:10.1073/pnas.0704772104, URL http://www.ncbi.nlm.nih.gov/pubmed/18003926.

Ahmed, T., Kilina, S., Das, T., Haraldsen, J. T., Rehr, J. J. and Balatsky, A. V. (2012). Electronic fingerprints of DNA bases on graphene, *Nano Lett* **12**, 2, pp. 927–931, doi:10.1021/nl2039315, URL http://www.ncbi.nlm.nih.gov/pubmed/22257137.

Alexander, C. and Sadiku, M. (2012). *Fundamentals of Electric Circuits: Fifth Edition* (McGraw-Hill Higher Education), ISBN 9780077575915, URL https://books.google.com.au/books?id=BUk1AAAAQBAJ.

Ambrosi, A., Chua, C. K., Bonanni, A. and Pumera, M. (2014). Electrochemistry of graphene and related materials, *Chem Rev* **114**, 14, pp. 7150–7188, doi:10.1021/cr500023c, URL http://www.ncbi.nlm.nih.gov/pubmed/24895834.

Antony, J. and Grimme, S. (2008). Structures and interaction energies of stacked graphene-nucleobase complexes, *Phys Chem Chem Phys* **10**, 19, p. 8, doi:10.1039/b805489b, URL http://www.ncbi.nlm.nih.gov/pubmed/18464972.

Ashcroft, N. W. and Mermin, N. D. (1976). *Solid State Physics* (Holt, New York), ISBN 0030839939, URL http://www.loc.gov/catdir/enhancements/fy1311/74009772-t.html.

Atkins, P. and de Paula, J. (2006). *Atkins' Physical Chemistry*, 8th edn. (Oxford University Press), ISBN 9780199543373, URL https://books.google.com.au/books?id=BV6cAQAAQBAJ.

Attard, P. (1995). Ion condensation in the electric double layer and the corresponding Poisson-Boltzmann effective surface charge, *J Phys Chem* **99**, pp. 14174–14181.

Bae, S., Kim, H., Lee, Y., Xu, X., Park, J.-S., Zheng, Y., Balakrishnan, J., Lei, T., Ri Kim, H., Song, Y. I., Kim, Y.-J., Kim, K. S., Özyilmaz, B., Ahn, J.-H., Hong, B. H. and Iijima, S. (2010). Roll-to-roll production of 30-inch graphene films for transparent electrodes, *Nat Nanotechnol* **5**, p. 574, doi:10.1038/nnano.2010.132, URL http://dx.doi.org/10.1038/nnano.2010.132.

Bard, A., Stratmann, M. and Licht, S. (2002). *Encyclopedia of Electrochemistry, Semiconductor Electrodes and Photoelectrochemistry* (Wiley), ISBN 9783527303984, URL https://books.google.com.au/books?id=TvPknQEACAAJ.

Bard, A. J. and Faulkner, L. R. (2000). *Electrochemical Methods: Fundamentals and Applications* (John Wiley & Sons, New York, USA), ISBN 0471043729 978-0471043720.

Bazant, M. Z., Storey, B. D. and Kornyshev, A. A. (2011). Double layer in ionic liquids: overscreening versus crowding, *Phys Rev Lett* **106**, 4, p. 046102, doi:10.1103/PhysRevLett.106.046102, URL http://www.ncbi.nlm.nih.gov/pubmed/21405339.

Bazant, M. Z., Thornton, K. and Ajdari, A. (2004). Diffuse-charge dynamics in electrochemical systems, *Phys Rev E* **70**, 2 Pt 1, p. 021506, doi:10.1103/PhysRevE.70.021506, URL http://www.ncbi.nlm.nih.gov/pubmed/15447495.

Berger, C., Song, Z., Li, T., Li, X., Ogbazghi, A. Y., Feng, R., Dai, Z., Marchenkov, A. N., Conrad, E. H., First, P. N. and de Heer, W. A. (2004). Ultrathin epitaxial graphite: 2D electron gas properties and a route toward graphene-based nanoelectronics, *J Phys Chem B* **108**, 52, pp. 19912–19916, doi:10.1021/jp040650f, URL http://dx.doi.org/10.1021/jp040650f.

Berland, K., Chakarova-Kack, S. D., Cooper, V. R., Langreth, D. C. and Schroder, E. (2011). A van der Waals density functional study of adenine on graphene: single-molecular adsorption and overlayer binding, *J Phys Condens Matter* **23**, 13, p. 135001, doi:10.1088/0953-8984/23/13/135001, URL http://www.ncbi.nlm.nih.gov/pubmed/21403239.

Bernier, N., Bocquet, F., Allouche, A., Saikaly, W., Brosset, C., Thibault, J. and Charaï, A. (2008). A methodology to optimize the quantification of sp2 carbon fraction from k edge eels spectra, *Journal of Electron Spectroscopy and Related Phenomena* **164**, 1–3, pp. 34–43, doi:10.1016/j.elspec.2008.04.006, URL http://www.sciencedirect.com/science/article/pii/S0368204808000285.

Bikerman, J. J. (1942). Xxxix. Structure and capacity of electrical double layer, *The London, Edinburgh, and Dublin Philosophical Magazine and Journal of Science* **33**, 220, pp. 384–397, doi: 10.1080/14786444208520813, URL http://dx.doi.org/10.1080/14786444208520813.

Bird, R., Stewart, W. and Lightfoot, E. (2007). *Transport Phenomena* (Wiley), ISBN 9780470115398, URL https://books.google.com.au/books?id=L5FnNlIaGfcC.

Bohinc, K., Kralj-Iglic, V. and Iglic, A. (2001). Thickness of electrical double layer. Effect of ion size, *Electrochim Acta* **46**, pp. 3033–3040.

Borukhov, I., Andelman, D. and Orland, H. (1997). Steric effects in electrolytes: a modified Poisson-Boltzmann equation, *Phys Rev Lett* **79**, 3, pp. 435–438.

Borukhov, I., Andelman, D. and Orland, H. (2000). Adsorption of large ions from an electrolyte solution: a modified Poisson–Boltzmann equation, *Electrochim Acta* **46**, pp. 221–229.

Bostwick, A., Ohta, T., Seyller, T., Horn, K. and Rotenberg, E. (2006). Quasiparticle dynamics in graphene, *Nat Phys* **3**, p. 36, doi:10.1038/nphys477, URL http://dx.doi.org/10.1038/nphys477.

Boyd, T. and Sanderson, J. (2003). *The Physics of Plasmas* (Cambridge University Press, Cambridge, UK), ISBN 9780521459129, URL http://books.google.com.au/books?id=bAmqvuGTUJ4C.

Britnell, L., Gorbachev, R. V., Jalil, R., Belle, B. D., Schedin, F., Mishchenko, A., Georgiou, T., Katsnelson, M. I., Eaves, L., Morozov, S. V., Peres, N. M., Leist, J., Geim, A. K., Novoselov, K. S. and Ponomarenko, L. A. (2012). Field-effect tunneling transistor based on vertical graphene heterostructures, *Science* **335**, 6071, pp. 947–950, doi:10.1126/science.1218461, URL http://www.ncbi.nlm.nih.gov/pubmed/22300848.

Brownson, D. A. C., Kampourisa, D. K. and Banks, C. E. (2012). Graphene electrochemistry: fundamental concepts through to prominent applications, *Chem Soc Rev* **41**, pp. 6944–6976.

Brydson, R., Zhili, Z. and Brown, A. (2008). *Revisiting the Determination of Carbon sp2/sp3 Ratios via Analysis of the EELS Carbon K-Edge* (Springer Berlin Heidelberg, Berlin, Heidelberg), ISBN 978-3-540-85156-1, pp. 357–358, doi:10.1007/978-3-540-85156-1_179, URL http://dx.doi.org/10.1007/978-3-540-85156-1_179.

Buljan, H., Jablan, M. and Soljačić. (2013). Graphene plasmonics: damping of plasmons in graphene, *Nat Photonics* **7**, 5, pp. 346–348, doi:10.1038/nphoton.2013.103.

Burden, R. and Faires, J. (2010). *Numerical Analysis* (Cengage Learning), ISBN 9780538733519, URL https://books.google.com.au/books?id=zXnSxY9G2JgC.

Carminati, R. (2013). A probe for graphene electronics, *Nat Nanotechnol* **8**, pp. 802–803.

Castro Neto, A. H., Peres, N. M. R., Novoselov, K. S. and Geim, A. K. (2009). The electronic properties of graphene, *Rev Mod Phys* **81**, 1, pp. 109–162, doi: 10.1103/RevModPhys.81.109.

Chakarova-Käck, S., Schröder, E., Lundqvist, B. and Langreth, D. (2006). Application of van der waals density functional to an extended system: adsorption of benzene and naphthalene on graphite, *Phys Rev Lett* **96**, 14, doi:10.1103/PhysRevLett.96.146107.

Chapman, D. L. (1913). LI. A contribution to the theory of electrocapillarity, *Philos Mag* **25**, 148, pp. 475–481, doi:10.1080/14786440408634187, URL http://dx.doi.org/10.1080/14786440408634187.

Chen, F., Xia, J., Ferry, D. K. and Tao, N. (2009a). Dielectric screening enhanced performance in graphene fet, *Nano Lett* **9**, 7, pp. 2571–2574.

Chen, F., Xia, J. and Tao, N. (2009b). Ionic screening of charged-impurity scattering in graphene, *Nano Lett* **9**, 4, pp. 1621–1625.

Chen, H., Müller, M. B., Gilmore, K. J., Wallace, G. G. and Li, D. (2008a). Mechanically strong, electrically conductive, and biocompatible graphene paper, *Adv Mater* **20**, 18, pp. 3557–3561.

Chen, J. H., Jang, C., Adam, S., Fuhrer, M. S., Williams, E. D. and Ishigami, M. (2008b). Charged-impurity scattering in graphene, *Nat Phys* **4**, 5, pp. 377–381, doi:10.1038/nphys935.

Chen, J. H., Jang, C., Xiao, S., Ishigami, M. and Fuhrer, M. S. (2008c). Intrinsic and extrinsic performance limits of graphene devices on SiO_2, *Nat Nanotechnol* **3**, 4, pp. 206–209, doi:10.1038/nnano.2008.58, URL http://www.ncbi.nlm.nih.gov/pubmed/18654504.

Chen, Z., Nai, J., Ma, H. and Li, Z. (2014). Nickel hydroxide nanocrystals-modified glassy carbon electrodes for sensitive l-histidine detection, *Electrochim Acta* **116**, pp. 258–262, doi:10.1016/j.electacta.2013.10.153, URL http://www.sciencedirect.com/science/article/pii/S0013468613021233.

Cohen-Karni, T., Qing, Q., Li, Q., Fang, Y. and Lieber, C. M. (2010). Graphene and nanowire transistors for cellular interfaces and electrical recording, *Nano Lett* **10**, 3, pp. 1098–1102, doi:10.1021/nl1002608, URL https://www.ncbi.nlm.nih.gov/pubmed/20136098.

Côté, R., Fouquet, J. P. and Luo, W. (2011). Biased bilayer graphene as a helical quantum hall ferromagnet, *Phys Rev B* **84**, 23, doi:10.1103/PhysRevB.84.235301.

Crassee, I., Levallois, J., Walter, A. L., Ostler, M., Bostwick, A., Rotenberg, E., Seyller, T., van der Marel, D. and Kuzmenko, A. B. (2010). Giant faraday rotation in single- and multilayer graphene, *Nat Phys* **7**, 1, pp. 48–51, doi:10.1038/nphys1816.

Crassee, I., Orlita, M., Potemski, M., Walter, A. L., Ostler, M., Seyller, T., Gaponenko, I., Chen, J. and Kuzmenko, A. B. (2012). Intrinsic terahertz plasmons and magnetoplasmons in large scale monolayer graphene, *Nano Lett* **12**, 5, pp. 2470–2474, doi:10.1021/nl300572y, URL http://www.ncbi.nlm.nih.gov/pubmed/22519967.

Dankerl, M., Hauf, M. V., Lippert, A., Hess, L. H., Birner, S., Sharp, I. D., Mahmood, A., Mallet, P., Veuillen, J.-Y., Stutzmann, M. and Garrido, J. A. (2010). Graphene solution-gated field-effect transistor array for sensing applications, *Adv Funct Mater* **20**, 18, pp. 3117–3124, doi:10.1002/adfm.201000724.

Das Sarma, S., Adam, S., Hwang, E. H. and Rossi, E. (2011). Electronic transport in two-dimensional graphene, *Rev Mod Phys* **83**, 2, pp. 407–470, doi:10.1103/RevModPhys.83.407.

Davies, J. H. (1998). *The Physics of Low-dimensional Semiconductors: An Introduction* (Cambridge University Press), ISBN 9780521484916, URL https://books.google.com.au/books?id=_YXoT2Tyq6oC.

Dion, M., Rydberg, H., Schröder, E., Langreth, D. C. and Lundqvist, B. I. (2004). Van der Waals density functional for general geometries, *Phys Rev Lett* **92**, 24, doi:10.1103/PhysRevLett.92.246401.

Dorgan, V. E., Bae, M.-H. and Pop, E. (2010). Mobility and saturation velocity in graphene on SiO_2, *Appl Phys Lett* **97**, 8, p. 082112, doi:10.1063/1.3483130.

Eigen, M. and Wick, E. (1954). The thermodynamics of electrolytes at higher concentration, *J Phys Chem* **58**, pp. 702–714.

El-Kady, M. F., Strong, V., Dubin, S. and Kaner, R. B. (2012). Laser scribing of high-performance and flexible graphene-based electrochemical capacitors, *Science* **335**, 6074, pp. 1326–1330, doi:10.1126/science.1216744, URL http://www.ncbi.nlm.nih.gov/pubmed/22422977.

Emtsev, K. V., Bostwick, A., Horn, K., Jobst, J., Kellogg, G. L., Ley, L., McChesney, J. L., Ohta, T., Reshanov, S. A., Röhrl, J., Rotenberg, E., Schmid, A. K., Waldmann, D., Weber, H. B. and Seyller, T. (2009). Towards wafer-size graphene layers by atmospheric pressure graphitization of silicon carbide, *Nat Mater* **8**, p. 203, doi:10.1038/nmat2382, URL http://dx.doi.org/10.1038/nmat2382.

Erickson, K., Erni, R., Lee, Z., Alem, N., Gannett, W. and Zettl, A. (2010). Determination of the local chemical structure of graphene oxide and reduced graphene oxide, *Adv Mater* **22**, 40, pp. 4467–4472, doi:10.1002/adma.201000732, URL http://dx.doi.org/10.1002/adma.201000732.

Farmer, D. B., Chiu, H.-Y., Lin, Y.-M., Jenkins, K. A., Xia, F. and Avouris, P. (2009). Utilization of a buffered dielectric to achieve high field-effect carrier mobility in graphene transistors, *Nano Lett* **9**, 12, p. 5.

Fei, Z., Rodin, A. S., Gannett, W., Dai, S., Regan, W., Wagner, M., Liu, M. K., McLeod, A. S., Dominguez, G., Thiemens, M., Castro Neto, A. H., Keilmann, F., Zettl, A., Hillenbrand, R., Fogler, M. M. and Basov, D. N. (2013). Electronic and plasmonic phenomena at graphene grain boundaries, *Nat Nanotechnol* **8**, 11, pp. 821–825, doi:10.1038/nnano.2013.197, URL http://www.ncbi.nlm.nih.gov/pubmed/24122082.

Ferrari, A. C. and Basko, D. M. (2013). Raman spectroscopy as a versatile tool for studying the properties of graphene, *Nat Nanotechnol* **8**, 4, pp. 235–246, doi:10.1038/nnano.2013.46, URL http://www.ncbi.nlm.nih.gov/pubmed/23552117.

Ferrari, A. C., Bonaccorso, F., Fal'ko, V., Novoselov, K. S., Roche, S., Boggild, P., Borini, S., Koppens, F. H. L., Palermo, V., Pugno, N., Garrido, J. A., Sordan, R., Bianco, A., Ballerini, L., Prato, M., Lidorikis, E., Kivioja, J., Marinelli, C., Ryhanen, T., Morpurgo, A., Coleman, J. N., Nicolosi, V., Colombo, L., Fert, A., Garcia-Hernandez, M., Bachtold, A., Schneider, G. F., Guinea, F., Dekker, C., Barbone, M., Sun, Z., Galiotis, C., Grigorenko, A. N., Konstantatos, G., Kis, A., Katsnelson, M., Vandersypen, L., Loiseau, A., Morandi, V., Neumaier, D., Treossi, E., Pellegrini, V., Polini, M., Tredicucci, A., Williams, G. M., Hee Hong, B., Ahn, J.-H., Min Kim, J., Zirath, H., van Wees, B. J., van der Zant, H., Occhipinti, L., Di Matteo, A., Kinloch, I. A., Seyller, T., Quesnel, E., Feng, X., Teo, K., Rupesinghe, N., Hakonen, P., Neil, S. R. T., Tannock, Q., Lofwander, T. and Kinaret, J. (2015). Science and technology roadmap for graphene, related two-dimensional crystals, and hybrid systems, *Nanoscale* **7**, 11, pp. 4598–4810, doi:10.1039/C4NR01600A, URL http://dx.doi.org/10.1039/C4NR01600A.

Ferrari, A. C., Meyer, J. C., Scardaci, V., Casiraghi, C., Lazzeri, M., Mauri, F., Piscanec, S., Jiang, D., Novoselov, K. S., Roth, S. and Geim, A. K. (2006). Raman spectrum of graphene and graphene layers, *Phys Rev Lett* **97**, 18, p. 187401, URL http://link.aps.org/doi/10.1103/PhysRevLett.97.187401.

Ferrari, A. C. and Robertson, J. (2000). Interpretation of raman spectra of disordered and amorphous carbon, *Phys Rev B* **61**, 20, pp. 14095–14107, URL http://link.aps.org/doi/10.1103/PhysRevB.61.14095.

Fiori, G., Bonaccorso, F., Iannaccone, G., Palacios, T., Neumaier, D., Seabaugh, A., Banerjee, S. K. and Colombo, L. (2014). Electronics based on two-dimensional materials, *Nat Nano* **9**, 10, pp. 768–779, doi:10.1038/nnano.2014.207, URL http://dx.doi.org/10.1038/nnano.2014.207.

Forest, L. D. (1906). The audion: a new receiver for wireless telegraphy, *Proc Am Inst Electr Eng* **25**, 10, pp. 719–747, doi:10.1109/PAIEE.1906.6741775.

Freitas, R. R., Rivelino, R., Mota Fde, B. and de Castilho, C. M. (2011). Dft studies of the interactions of a graphene layer with small water aggregates, *J Phys Chem A* **115**, 44, pp. 12348–12356, doi:10.1021/jp208279a, URL https://www.ncbi.nlm.nih.gov/pubmed/21954939.

Fu, W., Nef, C., Knopfmacher, O., Tarasov, A., Weiss, M., Calame, M. and Schonenberger, C. (2011). Graphene transistors are insensitive to pH changes in solution, *Nano Lett* **11**, 9, pp. 3597–3600, doi:10.1021/nl201332c, URL http://www.ncbi.nlm.nih.gov/pubmed/21766793.

Fu, W., Nef, C., Tarasov, A., Wipf, M., Stoop, R., Knopfmacher, O., Weiss, M., Calame, M. and Schonenberger, C. (2013). High mobility graphene ion-sensitive field-effect transistors by noncovalent functionalization, *Nanoscale* **5**, 24, pp. 12104–12110, doi:10.1039/c3nr03940d, URL https://www.ncbi.nlm.nih.gov/pubmed/24142362.

Garaj, S., Hubbard, W., Reina, A., Kong, J., Branton, D. and Golovchenko, J. A. (2010). Graphene as a subnanometre trans-electrode membrane, *Nature* **467**, 7312, pp. 190–193, doi:10.1038/nature09379, URL http://www.ncbi.nlm.nih.gov/pubmed/20720538.

Geim, A. K. (2009). Graphene: Status and prospects, *Science* **324**, 5934, pp. 1530–1534, doi:10.1126/science.1158877.

Geim, A. K. and Novoselov, K. S. (2007). The rise of graphene, *Nat Mater* **6**, pp. 183–191.

Gellings, P. and Bouwmeester, H. (1997). *Handbook of Solid State Electrochemistry* (Taylor & Francis), ISBN 9781420049305, URL https://books.google.com.au/books?id=SI07OAlf90oC.

Georgiou, T., Jalil, R., Belle, B. D., Britnell, L., Gorbachev, R. V., Morozov, S. V., Kim, Y. J., Gholinia, A., Haigh, S. J., Makarovsky, O., Eaves, L., Ponomarenko, L. A., Geim, A. K., Novoselov, K. S. and Mishchenko, A. (2013). Vertical field-effect transistor based on graphene-WS2 heterostructures for flexible and transparent electronics, *Nat Nanotechnol*

8, 2, pp. 100–103, doi:10.1038/nnano.2012.224, URL http://www.ncbi.nlm.nih.gov/pubmed/23263726.

Gerischer, H. (1990). The impact of semiconductors on the concepts of electrochemistry, *Electrochim Acta* **35**, 11–12, pp. 1677–1699, doi:http://dx.doi.org/10.1016/0013-4686(90)87067-C, URL //www.sciencedirect.com/science/article/pii/001346869087067C.

Gerischer, H. and Ekardt, W. (1983). Fermi levels in electrolytes and the absolute scale of redox potentials, *Appl Phys Lett* **43**, 4, pp. 393–395, doi:10.1063/1.94356, URL http://aip.scitation.org/doi/abs/10.1063/1.94356.

Gouy, G. (1917). Sur la fonction électrocapillaire, *Ann Phys* **9**, 7, pp. 129–184, doi:10.1051/anphys/191709070129, URL https://doi.org/10.1051/anphys/191709070129.

Gouy, M. (1910). Sur la constitution de la charge électrique à la surface d'un électrolyte, *J Phys Theor Appl* **9**, 1, pp. 457–468.

Gowtham, S., Scheicher, R., Ahuja, R., Pandey, R. and Karna, S. (2007). Physisorption of nucleobases on graphene: density-functional calculations, *Phys Rev B* **76**, 3, doi:10.1103/PhysRevB.76.033401.

Grabert, H., Devoret, M. and Division, N. A. T. O. S. A. (1992). *Single Charge Tunneling: Coulomb Blockade Phenomena in Nanostructures* (Springer), ISBN 9780306442292, URL https://books.google.com.au/books?id=5mEsAAAAYAAJ.

Gressman, P. T. and Strain, R. M. (2010). Global classical solutions of the boltzmann equation with long-range interactions, *Proc Natl Acad Sci U S A* **107**, 13, pp. 5744–5749, doi:10.1073/pnas.1001185107, URL https://www.ncbi.nlm.nih.gov/pubmed/20231489.

Grigorenko, A. N., Polini, M. and Novoselov, K. S. (2012). Graphene plasmonics, *Nat Photonics* **6**, 11, pp. 749–758, doi:10.1038/nphoton.2012.262.

Gu, N., Rudner, M., Young, A., Kim, P. and Levitov, L. (2011). Collapse of Landau levels in gated graphene structures, *Phys Rev Lett* **106**, 6, doi:10.1103/PhysRevLett.106.066601.

Gusynin, V. P. and Sharapov, S. G. (2005). Unconventional integer quantum hall effect in graphene, *Phys Rev Lett* **95**, 14, p. 146801, doi:10.1103/PhysRevLett.95.146801, URL https://www.ncbi.nlm.nih.gov/pubmed/16241680.

Hammouri, M., Jha, S. K. and Vasiliev, I. (2015). First-principles study of graphene and carbon nanotubes functionalized with benzyne, *J Phys*

Chem C **119**, 32, pp. 18719–18728, doi:10.1021/acs.jpcc.5b04065, URL http://dx.doi.org/10.1021/acs.jpcc.5b04065.

Hao, Y., Bharathi, M. S., Wang, L., Liu, Y., Chen, H., Nie, S., Wang, X., Chou, H., Tan, C., Fallahazad, B., Ramanarayan, H., Magnuson, C. W., Tutuc, E., Yakobson, B. I., McCarty, K. F., Zhang, Y.-W., Kim, P., Hone, J., Colombo, L. and Ruoff, R. S. (2013). The role of surface oxygen in the growth of large single-crystal graphene on copper, *Science* **342**, 6159, pp. 720–723, doi: 10.1126/science.1243879.

Hassibi, A., Navid, R., Dutton, R. W. and Lee, T. H. (2004). Comprehensive study of noise processes in electrode electrolyte interfaces, *J Appl Phys* **96**, 2, pp. 1074–1082, doi:10.1063/1.1755429, URL https://doi.org/10.1063/1.1755429.

Heller, I., Chatoor, S., Mannik, J., Zevenbergen, M. A., Oostinga, J. B., Morpurgo, A. F., Dekker, C. and Lemay, S. G. (2010). Charge noise in graphene transistors, *Nano Lett* **10**, 5, pp. 1563–1567, doi:10.1021/nl903665g, URL https://www.ncbi.nlm.nih.gov/pubmed/20373788.

Hess, L. H., Seifert, M. and Garrido, J. A. (2013). Graphene transistors for bioelectronics, *Proc IEEE* **101**, 7, pp. 1780–1792, doi:10.1109/JPROC.2013.2261031.

Hoggard, A., Wang, L.-Y., Ma, L., Fang, Y., You, G., Olson, J., Liu, Z., Chang, W.-S., Ajayan, P. M. and Link, S. (2013). Using the plasmon linewidth to calculate the time and efficiency of electron transfer between gold nanorods and graphene, *ACS Nano* **7**, 12, p. 9.

Huang, P. Y., Ruiz-Vargas, C. S., van der Zande, A. M., Whitney, W. S., Levendorf, M. P., Kevek, J. W., Garg, S., Alden, J. S., Hustedt, C. J., Zhu, Y., Park, J., McEuen, P. L. and Muller, D. A. (2011). Grains and grain boundaries in single-layer graphene atomic patchwork quilts, *Nature* **469**, p. 389, doi: 10.1038/nature09718, URL http://dx.doi.org/10.1038/nature09718.

Hummers, W. S. and Offeman, R. E. (1958). Preparation of graphitic oxide, *J Am Chem Soc* **80**, 6, pp. 1339–1339, doi:10.1021/ja01539a017, URL http://dx.doi.org/10.1021/ja01539a017.

Hwang, E., Adam, S. and Sarma, S. (2007a). Carrier transport in two-dimensional graphene layers, *Phys Rev Lett* **98**, 18, doi:10.1103/PhysRevLett.98.186806.

Hwang, E. H., Adam, S. and Das Sarma, S. (2007b). Transport in chemically doped graphene in the presence of adsorbed molecules, *Phys Rev B* **76**, p. 195421.

Ionescu, A. M. and Riel, H. (2011). Tunnel field-effect transistors as energy-efficient electronic switches, *Nature* **479**, 7373, pp. 329–337, URL http://dx.doi.org/10.1038/nature10679.

Islam, M. M., Alam, M. T. and Ohsaka, T. (2008). Electrical double-layer structure in ionic liquids: a corroboration of the theoretical model by experimental results, *J Phys Chem C* **112**, 42, pp. 16568–16574, doi:10.1021/jp8058849, URL http://dx.doi.org/10.1021/jp8058849.

Jackson, J. (2007). *Classical Electrodynamics* (Wiley), ISBN 9788126510948, URL https://books.google.com.au/books?id=8qHCZjJHRUgC.

Jang, C., Adam, S., Chen, J. H., Williams, E. D., Das Sarma, S. and Fuhrer, M. S. (2008). Tuning the effective fine structure constant in graphene: Opposing effects of dielectric screening on short- and long-range potential scattering, *Phys Rev Lett* **101**, 14, doi:10.1103/PhysRevLett.101.146805.

Jena, D. and Konar, A. (2007). Enhancement of carrier mobility in semiconductor nanostructures by dielectric engineering, *Phys Rev Lett* **98**, 13, p. 136805, doi:10.1103/PhysRevLett.98.136805.

Ji, D., Liu, L., Li, S., Chen, C., Lu, Y., Wu, J. and Liu, Q. (2017). Smartphone-based cyclic voltammetry system with graphene modified screen printed electrodes for glucose detection, *Biosens Bioelectron* **98**, Suppl C, pp. 449–456, doi:10.1016/j.bios.2017.07.027, URL http://www.sciencedirect.com/science/article/pii/S0956566317304748.

Ji, H., Zhao, X., Qiao, Z., Jung, J., Zhu, Y., Lu, Y., Zhang, L. L., MacDonald, A. H. and Ruoff, R. S. (2014). Capacitance of carbon-based electrical double-layer capacitors, *Nat Commun* **5**, p. 3317, doi:10.1038/ncomms4317, URL http://www.ncbi.nlm.nih.gov/pubmed/24557361.

Ju, L., Geng, B., Horng, J., Girit, C., Martin, M., Hao, Z., Bechtel, H. A., Liang, X., Zettl, A., Shen, Y. R. and Wang, F. (2011). Graphene plasmonics for tunable terahertz metamaterials, *Nat Nanotechnol* **6**, 10, pp. 630–634, doi:10.1038/nnano.2011.146, URL http://www.ncbi.nlm.nih.gov/pubmed/21892164.

Kilic, M. S., Bazant, M. Z. and Ajdari, A. (2007). Steric effects in the dynamics of electrolytes at large applied voltages. I. Double-layer charging, *Phys Rev E* **75**, 2 Pt 1, p. 021502, doi:10.1103/PhysRevE.75.021502, URL http://www.ncbi.nlm.nih.gov/pubmed/17358343.

Kim, C.-H. and Frisbie, C. D. (2014). Determination of quantum capacitance and band filling potential in graphene transistors with dual electrochemical and field-effect gates, *J Phys Chem C* **118**, 36, pp. 21160–21169, doi:10.1021/jp505391u.

Kim, Y. R., Bong, S., Kang, Y. J., Yang, Y., Mahajan, R. K., Kim, J. S. and Kim, H. (2010). Electrochemical detection of dopamine in the presence of ascorbic acid using graphene modified electrodes, *Biosens Bioelectron* **25**, 10, pp. 2366–2369, doi:10.1016/j.bios.2010.02.031, URL http://www.ncbi.nlm.nih.gov/pubmed/20307965.

Kittel, C. (2004). *Introduction to Solid State Physics* (Wiley), ISBN 9780471415268, URL https://books.google.com.au/books?id=kym4QgAACAAJ.

Knight, D. S. and White, W. B. (1988). Characterization of diamond films by raman spectroscopy, *J Mater Res* **4**, 2, pp. 385–393.

Kondrat, S. and Kornyshev, A. (2011). Superionic state in double-layer capacitors with nanoporous electrodes, *J Phys: Condens Matter* **23**, 2, p. 022201, URL http://stacks.iop.org/0953-8984/23/i=2/a=022201.

Kong, L., Enders, A., Rahman, T. S. and Dowben, P. A. (2014). Molecular adsorption on graphene, *J Phys: Condens Matter* **26**, 44, p. 443001, doi:10.1088/0953-8984/26/44/443001, URL http://www.ncbi.nlm.nih.gov/pubmed/25287516.

Kornyshev, A. A. (2007). Double-layer in ionic liquids: paradigm change? *J Phys Chem B* **111**, pp. 5545–5557.

Kostarelos, K. and Novoselov, K. S. (2014a). Exploring the interface of graphene and biology, *Science* **344**, 6181, pp. 261–263, doi:10.1126/science.1246736, URL http://science.sciencemag.org/content/sci/344/6181/261.full.pdf.

Kostarelos, K. and Novoselov, K. S. (2014b). Graphene devices for life, *Nat Nano* **9**, 10, pp. 744–745, doi:10.1038/nnano.2014.224, URL http://dx.doi.org/10.1038/nnano.2014.224.

Kühne, M., Paolucci, F., Popovic, J., Ostrovsky, P. M., Maier, J. and Smet, J. H. (2017). Ultrafast lithium diffusion in bilayer graphene, *Nat Nanotechnol* **12**, p. 895, doi:10.1038/nnano.2017.108, URL http://dx.doi.org/10.1038/nnano.2017.108.

Kumar, B., Min, K., Bashirzadeh, M., Farimani, A. B., Bae, M. H., Estrada, D., Kim, Y. D., Yasaei, P., Park, Y. D., Pop, E., Aluru, N. R. and Salehi-Khojin, A. (2013). The role of external defects in chemical sensing of graphene field-effect transistors, *Nano Lett* **13**, 5, pp. 1962–1968, doi:10.1021/nl304734g, URL http://www.ncbi.nlm.nih.gov/pubmed/23586702.

Kundu, P. and Cohen, I. (2010). *Fluid Mechanics* (Elsevier Science), ISBN 9780123814005, URL https://books.google.com.au/books?id=d9B5NElxUKwC.

Lasia, A. (2014). *Electrochemical Impedance Spectroscopy and its Applications* (Springer New York), ISBN 9781461489320, URL https://books.google.com.au/books?id=hFzgoAEACAAJ.

Le, D., Kara, A., Schroder, E., Hyldgaard, P. and Rahman, T. S. (2012). Physisorption of nucleobases on graphene: a comparative van der waals study, *J Phys: Condens Matter* **24**, 42, p. 424210, doi:10.1088/0953-8984/24/42/424210, URL http://www.ncbi.nlm.nih.gov/pubmed/23032709.

Lee, C., Wei, X., Kysar, J. W. and Hone, J. (2008). Measurement of the elastic properties and intrinsic strength of monolayer graphene, *Science* **321**, 5887, pp. 385–388, doi:10.1126/science.1157996, URL http://www.ncbi.nlm.nih.gov/pubmed/18635798.

Lee, G. H., Cooper, R. C., An, S. J., Lee, S., van der Zande, A., Petrone, N., Hammerberg, A. G., Lee, C., Crawford, B., Oliver, W., Kysar, J. W. and Hone, J. (2013a). High-strength chemical-vapor-deposited graphene and grain boundaries, *Science* **340**, 6136, pp. 1073–1076, doi:10.1126/science.1235126, URL http://www.ncbi.nlm.nih.gov/pubmed/23723231.

Lee, J.-H., Choi, Y.-K., Kim, H.-J., Scheicher, R. H. and Cho, J.-H. (2013b). Physisorption of DNA nucleobases on h-BN and graphene: vdW-corrected DFT calculations, *J Phys Chem C* **117**, 26, pp. 13435–13441, doi:10.1021/jp402403f.

Leenaerts, O., Partoens, B. and Peeters, F. M. (2008). Adsorption of H_2O, NH_3, CO, NO_2, and NO on graphene: a first-principles study, *Phys Rev B* **77**, 12, p. 125416, doi:10.1103/PhysRevB.77.125416.

Leenaerts, O., Partoens, B. and Peeters, F. M. (2009). Water on graphene: hydrophobicity and dipole moment using density functional theory, *Phys Rev B* **79**, 23, p. 235440, doi:10.1103/PhysRevB.79.235440.

Levin, Y. (2002). Electrostatic correlations: from plasma to biology, *Rep Prog Phys* **65**, 11, p. 1577.

Lherbier, A., Blase, X., Niquet, Y.-M., Triozon, F. and Roche, S. (2008). Charge transport in chemically doped 2D graphene, *Phys Rev Lett* **101**, 3, p. 036808, doi:10.1103/PhysRevLett.101.036808.

Li, D., Muller, M. B., Gilje, S., Kaner, R. B. and Wallace, G. G. (2008). Processable aqueous dispersions of graphene nanosheets, *Nat Nano* **3**, 2, pp. 101–105, doi:10.1038/nnano.2007.451, URL http://dx.doi.org/10.1038/nnano.2007.451.

Li, H., Ying, H., Chen, X., Nika, D. L., Cocemasov, A. I., Cai, W., Balandin, A. A. and Chen, S. (2014a). Thermal conductivity of twisted bilayer graphene,

Nanoscale **6**, 22, pp. 13402–13408, doi:10.1039/C4NR04455J, URL http://dx.doi.org/10.1039/C4NR04455J.

Li, X., Cai, W., An, J., Kim, S., Nah, J., Yang, D., Piner, R., Velamakanni, A., Jung, I., Tutuc, E., Banerjee, S. K., Colombo, L. and Ruoff, R. S. (2009). Large-area synthesis of high-quality and uniform graphene films on copper foils, *Science* **324**, 5932, pp. 1312–1314, doi:10.1126/science.1171245.

Li, Y., Yan, H., Farmer, D. B., Meng, X., Zhu, W., Osgood, R. M., Heinz, T. F. and Avouris, P. (2014b). Graphene plasmon enhanced vibrational sensing of surface-adsorbed layers, *Nano Lett* **14**, 3, pp. 1573–1577, doi:10.1021/nl404824w, URL http://www.ncbi.nlm.nih.gov/pubmed/24528250.

Liang, J., Chen, Z., Guo, L. and Li, L. (2011). Electrochemical sensing of l-histidine based on structure-switching dnazymes and gold nanoparticle-graphene nanosheet composites, *Chem Commun* **47**, 19, pp. 5476–5478, doi:10.1039/C1CC10965K, URL http://dx.doi.org/10.1039/C1CC10965K.

Lin, M.-Y., Guo, W.-C., Wu, M.-H., Wang, P.-Y., Liu, T.-H., Pao, C.-W., Chang, C.-C., Lee, S.-C. and Lin, S.-Y. (2012). Low-temperature grown graphene films by using molecular beam epitaxy, *Appl Phys Lett* **101**, 22, p. 221911, doi:10.1063/1.4768948, URL https://doi.org/10.1063/1.4768948.

Lin, Y.-M., Dimitrakopoulos, C., Jenkins, K. A., Farmer, D. B., Chiu, H.-Y., Grill, A. and Avouris, P. (2010). 100-ghz transistors from wafer-scale epitaxial graphene, *Science* **327**, p. 1.

Lin, Y.-M., Jenkins, K. A., Valdes-Garcia, A., Small, J. P., Farmer, D. B. and Avouris, P. (2009). Operation of graphene transistors at gigahertz frequencies, *Nano Lett* **9**, 1, p. 5.

Lipschutz, S., Spiegel, M. and Spellman, D. (2009). *Vector Analysis and an Introduction to Tensor Analysis* (McGraw-Hill), ISBN 9780071615457, URL https://books.google.com.au/books?id=dUOwbxd07HIC.

Liu, M., Yin, X., Ulin-Avila, E., Geng, B., Zentgraf, T., Ju, L., Wang, F. and Zhang, X. (2011). A graphene-based broadband optical modulator, *Nature* **474**, 7349, pp. 64–67, doi:10.1038/nature10067, URL http://www.ncbi.nlm.nih.gov/pubmed/21552277.

Liu, Q., Zhu, X., Huo, Z., He, X., Liang, Y. and Xu, M. (2012). Electrochemical detection of dopamine in the presence of ascorbic acid using pvp/graphene modified electrodes, *Talanta* **97**, pp. 557–562, doi:10.1016/j.talanta.2012.05.013, URL http://www.ncbi.nlm.nih.gov/pubmed/22841122.

Lockett, V., Sedev, R., Ralston, J., Horne, M. and Rodopoulos, T. (2008). Differential capacitance of the electrical double layer in imidazolium-based ionic liquids: Influence of potential, cation size, and temperature, *J Phys Chem C* **112**, 19, pp. 7486–7495, doi:10.1021/jp7100732, URL http://dx.doi.org/10.1021/jp7100732.

Lu, J., Zhang, H., Shi, W., Wang, Z., Zheng, Y., Zhang, T., Wang, N., Tang, Z. and Sheng, P. (2011). Graphene magnetoresistance device in van der Pauw geometry, *Nano Lett* **11**, 7, pp. 2973–2977, doi:10.1021/nl201538m, URL http://www.ncbi.nlm.nih.gov/pubmed/21696168.

Lundstrom, M. (2009). *Fundamentals of Carrier Transport* (Cambridge University Press), ISBN 9780521637244, URL https://books.google.com.au/books?id=Vrkd_dSC3zwC.

Luryi, S. (1988). Quantum capacitance devices, *Appl Phys Lett* **52**, 6, pp. 501–503, doi:10.1063/1.99649, URL http://aip.scitation.org/doi/abs/10.1063/1.99649.

Marcus, R. A. (1956). On the theory of oxidation-reduction reactions involving electron transfer. I, *J Chem Phys* **24**, 5, pp. 966–978, doi:10.1063/1.1742723, URL http://scitation.aip.org/content/aip/journal/jcp/24/5/10.1063/1.1742723.

Martin, J., Akerman, N., Ulbricht, G., Lohmann, T., Smet, J. H., von Klitzing, K. and Yacoby, A. (2008). Observation of electron-hole puddles in graphene using a scanning single-electron transistor, *Nat Phys* **4**, 2, pp. 144–148, doi:10.1038/nphys781, URL http://dx.doi.org/10.1038/nphys781.

Matsumoto, M., Saito, Y., Park, C., Fukushima, T. and Aida, T. (2015). Ultrahigh-throughput exfoliation of graphite into pristine 'single-layer' graphene using microwaves and molecularly engineered ionic liquids, *Nat Chem* **7**, p. 730, doi:10.1038/nchem.2315, URL http://dx.doi.org/10.1038/nchem.2315.

Matthews, M. J., Pimenta, M. A., Dresselhaus, G., Dresselhaus, M. S. and Endo, M. (1999). Origin of dispersive effects of the Raman d band in carbon materials, *Phys Rev B* **59**, 10, pp. R6585–R6588, URL http://link.aps.org/doi/10.1103/PhysRevB.59.R6585.

McAllister, M. J., Li, J.-L., Adamson, D. H., Schniepp, H. C., Abdala, A. A., Liu, J., Herrera-Alonso, M., Milius, D. L., Car, R., Prud'homme, R. K. and Aksay, I. A. (2007). Single sheet functionalized graphene by oxidation and thermal expansion of graphite, *Chem Mater* **19**, 18, pp. 4396–4404, doi:10.1021/cm0630800, URL http://dx.doi.org/10.1021/cm0630800.

McCreery, R. L. (2008). Advanced carbon electrode materials for molecular electrochemistry, *Chem Rev* **108**, p. 42.

Memming, R. (2015). *Semiconductor Electrochemistry* (Wiley), ISBN 9783527688708, URL https://books.google.com.au/books?id=A6sxBwAAQBAJ.

Mezger, M., Schröder, H., Reichert, H., Schramm, S., Okasinski, J. S., Schöder, S., Honkimäki, V., Deutsch, M., Ocko, B. M., Ralston, J., Rohwerder, M., Stratmann, M. and Dosch, H. (2008). Molecular layering of fluorinated ionic liquids at a charged sapphire (0001) surface, *Science* **322**, 5900, pp. 424–428, doi:10.1126/science.1164502, URL http://science.sciencemag.org/content/sci/322/5900/424.full.pdf.

Min, S. K., Kim, W. Y., Cho, Y. and Kim, K. S. (2011). Fast DNA sequencing with a graphene-based nanochannel device, *Nat Nanotechnol* **6**, 3, pp. 162–165, doi:10.1038/nnano.2010.283, URL http://www.ncbi.nlm.nih.gov/pubmed/21297626.

Mkhoyan, K. A., Contryman, A. W., Silcox, J., Stewart, D. A., Eda, G., Mattevi, C., Miller, S. and Chhowalla, M. (2009). Atomic and electronic structure of graphene-oxide, *Nano Lett* **9**, 3, pp. 1058–1063, doi:10.1021/nl8034256, URL http://dx.doi.org/10.1021/nl8034256.

Moss, S. D., Janata, J. and Johnson, C. C. (1975). Potassium ion-sensitive field effect transistor, *Anal Chem* **47**, 13, pp. 2238–2243, doi:10.1021/ac60363a005, URL http://dx.doi.org/10.1021/ac60363a005.

Myamlin, V. A. and Yurii, V. P. (1963). The electrochemistry of semiconductors, *Russ Chem Rev* **32**, 4, p. 207, URL http://stacks.iop.org/0036-021X/32/i=4/a=R04.

Nair, R. R., Blake, P., Grigorenko, A. N., Novoselov, K. S., Booth, T. J., Stauber, T., Peres, N. M. R. and Geim, A. K. (2008). Fine structure constant defines visual transparency of graphene, *Science* **320**, p. 1308.

Nature Nanotechnology Feature (2014). Things you could do with graphene, *Nat Nano* **9**, 10, pp. 737–737, doi:10.1038/nnano.2014.245, URL http://dx.doi.org/10.1038/nnano.2014.245.

Newaz, A. K., Puzyrev, Y. S., Wang, B., Pantelides, S. T. and Bolotin, K. I. (2012). Probing charge scattering mechanisms in suspended graphene by varying its dielectric environment, *Nat Commun* **3**, p. 734, doi:10.1038/ncomms1740, URL http://www.ncbi.nlm.nih.gov/pubmed/22415825.

Nolte, D. D. (2010). The tangled tale of phase space, *Phys Today* **63**, 4, pp. 33–38, URL ⟨GotoISI⟩://WOS:000276548200021.

Nomura, K. and MacDonald, A. H. (2006). Quantum hall ferromagnetism in graphene, *Phys Rev Lett* **96**, 25, doi:10.1103/PhysRevLett.96.256602.

Novoselov, K. S., Geim, A. K., Morozov, S. V., Jiang, D., Katsnelson, M. I., Grigorieva, I. V., Dubonos, S. V. and Firsov, A. A. (2005). Two-dimensional gas of massless dirac fermions in graphene, *Nature* **438**, 7065, pp. 197–200, doi:10.1038/nature04233, URL http://www.ncbi.nlm.nih.gov/pubmed/16281030.

Novoselov, K. S., Geim, A. K., Morozov, S. V., Jiang, D., Zhang, Y., Dubonos, S. V., Grigorieva, I. V. and Firsov, A. A. (2004). Electric field effect in atomically thin carbon films, *Science* **306**, 5696, pp. 666–669, doi:10.1126/science.1102896, URL http://www.ncbi.nlm.nih.gov/pubmed/15499015.

Nyquist, H. (1928). Thermal agitation of electric charge in conductors, *Phys Rev* **32**, 1, pp. 110–113, URL https://link.aps.org/doi/10.1103/PhysRev.32.110.

Oldham, K. B. (2008). A gouy–chapman–stern model of the double layer at a (metal)/(ionic liquid) interface, *J Electroanal Chem* **613**, 2, pp. 131–138, doi:10.1016/j.jelechem.2007.10.017.

Ortmann, F., Schmidt, W. and Bechstedt, F. (2005). Attracted by long-range electron correlation: adenine on graphite, *Phys Rev Lett* **95**, 18, doi:10.1103/PhysRevLett.95.186101.

Pallecchi, E., Lafont, F., Cavaliere, V., Schopfer, F., Mailly, D., Poirier, W. and Ouerghi, A. (2014). High electron mobility in epitaxial graphene on 4h-SiC(0001) via post-growth annealing under hydrogen, *Sci Rep* **4**, p. 4558, doi:10.1038/srep04558, URL http://dx.doi.org/10.1038/srep04558.

Park, S., Floresca, H. C., Suh, Y. and Kim, M. J. (2010). Electron microscopy analyses of natural and highly oriented pyrolytic graphites and the mechanically exfoliated graphenes produced from them, *Carbon* **48**, 3, pp. 797–804, doi:10.1016/j.carbon.2009.10.030, URL http://www.sciencedirect.com/science/article/pii/S0008622309007039.

Pathria, R. and Beale, P. (2011). *Statistical Mechanics* (Elsevier Science), ISBN 9780123821898, URL https://books.google.com.au/books?id=KdbJJAXQ-RsC.

Perdew, J. P., Burke, K. and Ernzerhof, M. (1996). Generalized gradient approximation made simple, *Phys Rev Lett* **77**.

Perdew, J. P., Chevary, J. A., Vosko, S. H., Jackson, K. A., Pederson, M. R., Singh, D. J. and Fiolhais, C. (1992). Atoms, molecules, solids, and surfaces: Applications of the generalized gradient approximation for exchange and correlation, *Phys Rev B* **46**, 11, pp. 6671–6687, URL https://link.aps.org/doi/10.1103/PhysRevB.46.6671.

Perdew, J. P. and Zunger, A. (1981). Self-interaction correction to density-functional approximations for many-electron systems, *Phys Rev B* **23**,

10, pp. 5048–5079, URL https://link.aps.org/doi/10.1103/PhysRevB.23.5048.

Pi, K., Han, W., McCreary, K. M., Swartz, A. G., Li, Y. and Kawakami, R. K. (2010). Manipulation of spin transport in graphene by surface chemical doping, *Phys Rev Lett* **104**, 18, p. 187201, doi:10.1103/PhysRevLett.104.187201, URL https://www.ncbi.nlm.nih.gov/pubmed/20482203.

Ping, J., Yudhistira, I., Ramakrishnan, N., Cho, S., Adam, S. and Fuhrer, M. S. (2014). Disorder-induced magnetoresistance in a two-dimensional electron system, *Phys Rev Lett* **113**, 4, doi:10.1103/PhysRevLett.113.047206.

Ponomarenko, L. A., Yang, R., Gorbachev, R. V., Blake, P., Mayorov, A. S., Novoselov, K. S., Katsnelson, M. I. and Geim, A. K. (2010). Density of states and zero landau level probed through capacitance of graphene, *Phys Rev Lett* **105**, 13, p. 136801, doi:10.1103/PhysRevLett.105.136801, URL https://www.ncbi.nlm.nih.gov/pubmed/21230795.

Ponomarenko, L. A., Yang, R., Mohiuddin, T. M., Katsnelson, M. I., Novoselov, K. S., Morozov, S. V., Zhukov, A. A., Schedin, F., Hill, E. W. and Geim, A. K. (2009). Effect of a high- environment on charge carrier mobility in graphene, *Phys Rev Lett* **102**, 20, doi:10.1103/PhysRevLett.102.206603.

Pumera, M. (2013). Electrochemistry of graphene, graphene oxide and other graphenoids: review, *Electrochem Commun* **36**, pp. 14–18, doi:10.1016/j.elecom.2013.08.028.

Pumera, M., Ambrosi, A., Bonanni, A., Chng, E. L. K. and Poh, H. L. (2010). Graphene for electrochemical sensing and biosensing, *TrAC Trends Anal Chem* **29**, 9, pp. 954–965, doi:10.1016/j.trac.2010.05.011.

Sakurai, J. and Napolitano, J. (2011). *Modern Quantum Mechanics* (Addison-Wesley), ISBN 9780805382914, URL https://books.google.com.au/books?id=N4I-AQAACAAJ.

Schedin, F., Geim, A. K., Morozov, S. V., Hill, E. W., Blake, P., Katsnelson, M. I. and Novoselov, K. S. (2007). Detection of individual gas molecules adsorbed on graphene, *Nat Mater* **6**, 9, pp. 652–655, doi:10.1038/nmat1967, URL http://www.ncbi.nlm.nih.gov/pubmed/17660825.

Schwierz, F. (2010). Graphene transistors, *Nat Nano* **5**, 7, pp. 487–496, URL http://dx.doi.org/10.1038/nnano.2010.89.

Semiconductor Industry Association (2009). The International Technology Roadmap for Semiconductors, URL http://www.itrs.net/Links/2009ITRS/Home2009.htm, [www.itrs.net/Links/2009ITRS/Home2009.htm].

Shang, N. G., Papakonstantinou, P., McMullan, M., Chu, M., Stamboulis, A., Potenza, A., Dhesi, S. S. and Marchetto, H. (2008). Catalyst-free efficient growth, orientation and biosensing properties of multilayer graphene nanoflake films with sharp edge planes, *Adv Funct Mater* **18**, 21, pp. 3506–3514, doi:10.1002/adfm.200800951.

Shao, Y., Wang, J., Wu, H., Liu, J., Aksay, I. and Lin, Y. (2010). Graphene based electrochemical sensors and biosensors: a review, *Electroanalysis* **22**, 10, pp. 1027–1036, doi:10.1002/elan.200900571.

Siegel, D., Regan, W., Fedorov, A., Zettl, A. and Lanzara, A. (2013). Charge-carrier screening in single-layer graphene, *Phys Rev Lett* **110**, 14, doi: 10.1103/PhysRevLett.110.146802.

Simon, P. and Gogotsi, Y. (2008). Materials for electrochemical capacitors, *Nat Mater* **7**, 11, pp. 845–854, URL http://dx.doi.org/10.1038/nmat2297.

Singh, A. K., Uddin, M. A., Tolson, J. T., Maire-Afeli, H., Sbrockey, N., Tompa, G. S., Spencer, M. G., Vogt, T., Sudarshan, T. S. and Koley, G. (2013). Electrically tunable molecular doping of graphene, *Appl Phys Lett* **102**, 4, p. 043101, doi:10.1063/1.4789509.

Siwy, Z. S. and Davenport, M. (2010). Nanopores: graphene opens up to DNA, *Nat Nanotechnol* **5**, 10, pp. 697–698, doi:10.1038/nnano.2010.198, URL http://www.ncbi.nlm.nih.gov/pubmed/20924388.

Skinner, B., Loth, M. S. and Shklovskii, B. I. (2010). Capacitance of the double layer formed at the metal/ionic-conductor interface: how large can it be? *Phys Rev Lett* **104**, 12, p. 128302, URL https://link.aps.org/doi/10.1103/PhysRevLett.104.128302.

Stafiej, J., Ekoka, A., Borkowska, Z. and Badiali, J. P. (1996). New theoretical description of electrified interfaces, *J Chem Soc, Faraday Trans* **92**, 20, pp. 3677–3682, doi:10.1039/FT9969203677, URL http://dx.doi.org/10.1039/FT9969203677.

Stern, O. (1924). Zur theorie der elektrolytischen doppelschicht, *Z Elektrochem Angew Phys Chem* **30**, 21-22, pp. 508–516, doi:10.1002/bbpc.192400182, URL http://dx.doi.org/10.1002/bbpc.192400182.

Sun, J., Fertig, H. A. and Brey, L. (2010). Effective magnetic fields in graphene superlattices, *Phys Rev Lett* **105**, 15, doi:10.1103/PhysRevLett.105.156801.

Sze, S. (1981). *Physics of Semiconductor Devices* (John Wiley & Sons, Hoboken, New Jersey, USA), ISBN 9780471056614, URL http://books.google.com.au/books?id=LCNTAAAAMAAJ.

Szkopek, T. (2013). Nanoelectronic Devices.

Tiwari, R. P. and Stroud, D. (2009). Model for the magnetoresistance and hall coefficient of inhomogeneous graphene, *Phys Rev B* **79**, 16, doi: 10.1103/PhysRevB.79.165408.

Tkatchenko, A. and Scheffler, M. (2009). Accurate molecular van der waals interactions from ground-state electron density and free-atom reference data, *Phys Rev Lett* **102**, 7, doi:10.1103/PhysRevLett.102.073005.

Traversi, F., Raillon, C., Benameur, S. M., Liu, K., Khlybov, S., Tosun, M., Krasnozhon, D., Kis, A. and Radenovic, A. (2013). Detecting the translocation of DNA through a nanopore using graphene nanoribbons, *Nat Nanotechnol* **8**, 12, pp. 939–945, doi:10.1038/nnano.2013.240, URL http://www.ncbi.nlm.nih.gov/pubmed/24240429.

Tuinstra, F. and Koenig, J. L. (1970). Raman spectrum of graphite, *J Chem Phys* **53**, 3, pp. 1126–1130, doi:10.1063/1.1674108, URL http://dx.doi.org/10.1063/1.1674108.

Valette, G. (1982). Double layer on silver single crystal electrodes in contact with electrolytes having anions which are slightly specifically adsorbed part II. The (100) face, *J Electroanal Chem* **138**, pp. 37–54.

Wallace, P. R. (1947). The band theory of graphite, *Phys Rev* **71**, pp. 622–634, doi:10.1103/PhysRev.71.622, URL http://link.aps.org/doi/10.1103/PhysRev.71.622.

Wang, J., Zhu, M., Outlaw, R. A., Zhao, X., Manos, D. M. and Holloway, B. C. (2004). Synthesis of carbon nanosheets by inductively coupled radio-frequency plasma enhanced chemical vapor deposition, *Carbon* **42**, 14, pp. 2867–2872, doi:10.1016/j.carbon.2004.06.035, URL http://www.sciencedirect.com/science/article/pii/S000862230400435X.

Wang, Y., Li, Y., Tang, L., Lu, J. and Li, J. (2009). Application of graphene-modified electrode for selective detection of dopamine, *Electrochem Commun* **11**, 4, pp. 889–892, doi:10.1016/j.elecom.2009.02.013.

Wehling, T. O., Novoselov, K. S., Morozov, S. V., Vdovin, E. E., Katsnelson, M. I., Geim, A. K. and Lichtenstein, A. I. (2008). Molecular doping of graphene, *Nano Lett* **8**, 1, pp. 173–177.

Wu, Y., Lin, Y. M., Bol, A. A., Jenkins, K. A., Xia, F., Farmer, D. B., Zhu, Y. and Avouris, P. (2011). High-frequency, scaled graphene transistors on diamond-like carbon, *Nature* **472**, 7341, pp. 74–78, doi:10.1038/nature09979, URL http://www.ncbi.nlm.nih.gov/pubmed/21475197.

Xia, F., Farmer, D. B., Lin, Y. M. and Avouris, P. (2010a). Graphene field-effect transistors with high on/off current ratio and large transport band gap at room temperature, *Nano Lett* **10**, 2, pp. 715–718, doi:10.1021/nl9039636, URL http://www.ncbi.nlm.nih.gov/pubmed/20092332.

Xia, F., Yan, H. and Avouris, P. (2013). The interaction of lightvand graphene: Basics, devices, and applications, *Proc IEEE* **101**, 7, p. 15.

Xia, J., Chen, F., Li, J. and Tao, N. (2009). Measurement of the quantum capacitance of graphene, *Nat Nanotechnol* **4**, 8, pp. 505–509, doi:10.1038/nnano.2009.177, URL http://www.ncbi.nlm.nih.gov/pubmed/19662012.

Xia, J. L., Chen, F., Wiktor, P., Ferry, D. K. and Tao, N. J. (2010b). Effect of top dielectric medium on gate capacitance of graphene field effect transistors: implications in mobility measurements and sensor applications, *Nano Lett* **10**, 12, pp. 5060–5064, doi:10.1021/nl103306a, URL http://www.ncbi.nlm.nih.gov/pubmed/21090582.

Xiao, X., Beechem, T. E., Brumbach, M. T., Lambert, T. N., Davis, D. J., Michael, J. R., Washburn, C. M., Wang, J., Brozik, S. M., Wheeler, D. R., Burckel, D. B. and Polsky, R. (2012). Lithographically defined three-dimensional graphene structures, *ACS Nano* **6**, 4, pp. 3573–3579.

Xiong, G., Meng, C., Reifenberger, R. G., Irazoqui, P. P. and Fisher, T. S. (2014). A review of graphene-based electrochemical microsupercapacitors, *Electroanalysis* **26**, 1, pp. 30–51, doi:10.1002/elan.201300238, URL http://dx.doi.org/10.1002/elan.201300238.

Yan, F., Zhang, M. and Li, J. (2014). Solution-gated graphene transistors for chemical and biological sensors, *Adv Healthcare Mater* **3**, 3, pp. 313–331, doi:10.1002/adhm.201300221, URL http://dx.doi.org/10.1002/adhm.201300221.

Yan, H., Li, X., Chandra, B., Tulevski, G., Wu, Y., Freitag, M., Zhu, W., Avouris, P. and Xia, F. (2012). Tunable infrared plasmonic devices using graphene/insulator stacks, *Nat Nanotechnol* **7**, 5, pp. 330–334, doi:10.1038/nnano.2012.59, URL http://www.ncbi.nlm.nih.gov/pubmed/22522668.

Yan, H., Low, T., Zhu, W., Wu, Y., Freitag, M., Li, X., Guinea, F., Avouris, P. and Xia, F. (2013). Damping pathways of mid-infrared plasmons in graphene nanostructures, *Nat Photonics* **7**, 5, pp. 394–399, doi:10.1038/nphoton.2013.57.

Yang, X., Cheng, C., Wang, Y., Qiu, L. and Li, D. (2013). Liquid-mediated dense integration of graphene materials for compact capacitive energy storage, *Science* **341**, 6145, pp. 534–537, doi:10.1126/science.1239089, URL http://science.sciencemag.org.ezp.lib.unimelb.edu.au/content/sci/341/6145/534.full.pdf.

Yang, X., Zhu, J., Qiu, L. and Li, D. (2011). Bioinspired effective prevention of restacking in multilayered graphene films: towards the next generation of high-performance supercapacitors, *Adv Mater* **23**, 25, pp. 2833–2838, doi:10.1002/adma.201100261, URL http://dx.doi.org/10.1002/adma.201100261.

Ye, J., Craciun, M. F., Koshino, M., Russo, S., Inoue, S., Yuan, H., Shimotani, H., Morpurgo, A. F. and Iwasa, Y. (2011). Accessing the transport properties of graphene and its multilayers at high carrier density, *Proc Natl Acad Sci U S A* **108**, 32, pp. 13002–13006, doi:10.1073/pnas.1018388108, URL https://www.ncbi.nlm.nih.gov/pubmed/21828007.

Yu, Q., Lian, J., Siriponglert, S., Li, H., Chen, Y. P. and Pei, S.-S. (2008). Graphene segregated on ni surfaces and transferred to insulators, *Appl Phys Lett* **93**, 11, p. 113103, doi:10.1063/1.2982585, URL https://doi.org/10.1063/1.2982585.

Zeitler, R. and Fromherz, P. (2013). The thermal voltage fluctuations in the planar core-coat conductor of a neuron-semiconductor interface, *Langmuir* **29**, 20, pp. 6084–6090, doi:10.1021/la4002169, URL http://dx.doi.org/10.1021/la4002169.

Zhan, H., Cervenka, J., Prawer, S. and Garrett, D. (2018). Molecular detection by liquid gated Hall effect measurement of graphene, *Nanoscale* **10**, pp. 930–935, doi:10.1039/C7NR06330J, URL http://dx.doi.org/10.1039/C7NR06330J.

Zhan, H., Cervenka, J., Prawer, S. and Garrett, D. J. (2017). Electrical double layer at various electrode potentials: A modification by vibration, *J Phys Chem C* **121**, 8, pp. 4760–4764, doi:10.1021/acs.jpcc.7b00961, URL http://dx.doi.org/10.1021/acs.jpcc.7b00961.

Zhan, H., Garrett, D. J., Apollo, N. V., Ganesan, K., Lau, D., Prawer, S. and Cervenka, J. (2016). Direct fabrication of 3D graphene on nanoporous anodic alumina by plasma-enhanced chemical vapor deposition, *Sci Rep* **6**, p. 19822, doi:10.1038/srep19822, URL http://dx.doi.org/10.1038/srep19822.

Zhan, H. and Hu, C. (2011). Kinetic solutions for electrons in multi-cusp ion source, *Appl Phys Lett* **99**, 22, p. 221501, doi:10.1063/1.3664347, URL http://dx.doi.org/10.1063/1.3664347.

Zhan, H. L., Hu, C. D., Xie, Y. H., Wu, B., Wang, J. F., Liang, L. Z. and Wei, J. L. (2011a). Theoretical discussion for electron-density distribution in multicusp ion source, *Appl Phys Lett* **98**, 12, p. 121501, doi:10.1063/1.3571292, URL http://dx.doi.org/10.1063/1.3571292.

Zhan, N., Olmedo, M., Wang, G. and Liu, J. (2011b). Layer-by-layer synthesis of large-area graphene films by thermal cracker enhanced gas source molecular beam epitaxy, *Carbon* **49**, 6, pp. 2046–2052, doi: 10.1016/j.carbon.2011.01.033, URL http://www.sciencedirect.com/science/article/pii/S0008622311000522.

Zhang, F., Chen, X., Yu, C., Xu, X., Hu, X., Qin, X., Li, Q., Zhao, X., Yu, P. and Wang, R. (2017). High mobility and large domain decoupled epitaxial graphene on SiC (000$\bar{1}$) surface obtained by nearly balanced hydrogen etching, *Mater Lett* **195**, Suppl C, pp. 82–85, doi:10.1016/j.matlet.2017.02.105, URL http://www.sciencedirect.com/science/article/pii/S0167577X17302938.

Zhang, M., Liao, C., Mak, C. H., You, P., Mak, C. L. and Yan, F. (2015). Highly sensitive glucose sensors based on enzyme-modified whole-graphene solution-gated transistors, *Sci Rep* **5**, p. 8311, doi:10.1038/srep08311, URL http://dx. doi.org/ 10.1038/srep08311.

Zhang, Y., Tan, Y. W., Stormer, H. L. and Kim, P. (2005). Experimental observation of the quantum hall effect and berry's phase in graphene, *Nature* **438**, 7065, pp. 201–204, doi:10.1038/nature04235, URL https://www.ncbi.nlm.nih.gov/pubmed/16281031.

Zheng, Y. and Ando, T. (2002). Hall conductivity of a two-dimensional graphite system, *Phys Rev B* **65**, 24, doi:10.1103/PhysRevB.65.245420.

Zhu, X., Sun, L., Chen, Y., Ye, Z., Shen, Z. and Li, G. (2013). Combination of cascade chemical reactions with graphene-DNA interaction to develop new strategy for biosensor fabrication, *Biosens Bioelectron* **47**, pp. 32–37, doi:10.1016/j.bios.2013.02.039, URL http://www.ncbi.nlm.nih.gov/pubmed/23542067.

Index